THE
SCENERY OF SIDMOUTH

SALCOMBE REGIS

THE
SCENERY OF SIDMOUTH

Its Natural Beauty and Historic Interest

by

VAUGHAN CORNISH, D.Sc.

With Map and illustrations

CAMBRIDGE
AT THE UNIVERSITY PRESS
1940

BY THE SAME AUTHOR

The Poetic Impression of Natural Scenery
SIFTON, PRAED AND CO. LTD. (1931)

Scenery and the Sense of Sight
CAMBRIDGE UNIVERSITY PRESS (1934)

The Preservation of our Scenery
CAMBRIDGE UNIVERSITY PRESS (1937)

The Scenery of England (2nd ed.)
A. & C. BLACK, LTD. (1937)

CAMBRIDGE
UNIVERSITY PRESS

University Printing House, Cambridge CB2 8BS, United Kingdom

Cambridge University Press is part of the University of Cambridge.

It furthers the University's mission by disseminating knowledge in the pursuit of education, learning and research at the highest international levels of excellence.

www.cambridge.org
Information on this title: www.cambridge.org/9781107492783

© Cambridge University Press 1940

This publication is in copyright. Subject to statutory exception and to the provisions of relevant collective licensing agreements, no reproduction of any part may take place without the written permission of Cambridge University Press.

First published 1940
First paperback edition 2015

A catalogue record for this publication is available from the British Library

ISBN 978-1-107-49278-3 Paperback

Cambridge University Press has no responsibility for the persistence or accuracy of URLs for external or third-party internet websites referred to in this publication, and does not guarantee that any content on such websites is, or will remain, accurate or appropriate.

TO

SIR LAURENCE WENSLEY CHUBB, *Kt.*
Secretary of the Commons, Open Spaces and
Footpaths Preservation Society

With Esteem and Regard
I dedicate this book
V. C.

CONTENTS

Preface p. xi

Acknowledgments p. xiii

CHAPTER I. SIDMOUTH PARISH

1. The situation of Sidmouth. 2. The early days of Sidmouth as a seaside resort. 3. The panoramic picture of Sidmouth. 4. The villas and their gardens pp. 1–16

CHAPTER II. SIDMOUTH PARISH (*concluded*)

5. The coast as seen from the esplanade. 6. The sea and sky from the esplanade. 7. The geological investigation of the Sidmouth cliffs and combes. 8. Peak Hill and the westward outlook from the Sidmouth District. 9. Sidmouth Gap pp. 17–37

CHAPTER III. SIDBURY

10. Sidbury Castle. 11. Sidbury Manor Park and Sand Barton. 12. Upper Sid Vale and the Plateau pp. 38–48

CHAPTER IV. SALCOMBE REGIS

13. The Byes. 14. The Salcombe Regis Thorn. 15. Salcombe Regis Valley and the reservation of Thorn Farm as agricultural land. pp. 49–63

CHAPTER V. SALCOMBE REGIS (*concluded*)

16. The view from the summit of Salcombe Cliff. 17. The reservation of South Combe Farm on the Cliff as an Open Space. 18. Dunscombe Hill and Weston Combe pp. 64–81

Index pp. 83–86

ILLUSTRATIONS

1. Salcombe Regis   *frontispiece*
   The head of the combe where the heights converge towards the Thorn Tree.
   (*A photograph from 'The Times'*)

2. Map of Sidmouth Urban District   *facing page* 8

3. Sidmouth Gap from the Sea   34
   (*Sketch by the Author*)

4. The Salcombe Regis Thorn   56
   (*Sketch by R. W. Sampson, architect*)

5. The Cliff Fields of South Combe   76
   (*Photograph by the Author*)

6. Panorama of Sidmouth in 1815   *in folder*
   (*By Hubert Cornish*)

PREFACE

The movement for the preservation of Rural England depends largely for its success upon wider dissemination of the knowledge and appreciation of scenery.

Having endeavoured in former books and essays to establish general principles for the aesthetics of scenery, I decided to illustrate their application by the analytical description of a small area, every part of which could easily be viewed by resident or visitor.

The place which I have selected for this purpose is Sidmouth. Since I first knew the district in the days of childhood I have been a pilgrim of scenery in many lands, but no aspect of the world from Arctic to Equator has diminished my admiration for the beauty of this peaceful spot on the south coast of Devon.

Moreover the inheritance of a property of natural beauty and historic interest in this locality has enabled me, though in quite a small way, to put into practice the doctrines of the Council for the Preservation of Rural England.

VAUGHAN CORNISH

Inglewood, Gordon Road
Camberley, Surrey
January 15th, 1940

ACKNOWLEDGMENTS

The author returns thanks for information received from the Admiralty (Hydrographic Department), L. M. Blanchard (Surveyor, Sidmouth Urban District Council), the Very Rev. S. C. Carpenter (Dean of Exeter), D. L. Edwards, F.R.A.S. (Director, Norman Lockyer Observatory), Sir Richard Gregory, Bart., F.R.S., the Ordnance Survey Department, R. Pickard (Clerk, Sidmouth Urban District Council), the Ven. Lonsdale Ragg, D.D. and Heywood Sumner, F.S.A.

For permission to reproduce illustrations thanks are due to Miss Kestell-Cornish, Mr R. W. Sampson and *The Times*.

CHAPTER I

SIDMOUTH PARISH

§ 1. Sidmouth is situated at the head of the largest bay on the coast of the English Channel, extending from the Isle of Portland to Start Point. West of Lyme Regis, where Dorset passes into Devon, the coast is a line of cliffs, and until we come to the estuary of the Exe there are only three gaps in the cliff rampart which are of sufficient width to favour the growth of seaside resorts. In these gaps stand Seaton, Sidmouth and Budleigh Salterton. The first gap is that made by the River Axe, and here is Seaton. The last gap is that made by the River Otter. On the west side of the mouth of this river is Budleigh Salterton.

The Otter and the Yarty, the western tributary, or, more properly, component of the River Axe, rise close to one another, 16 miles from the coast, under Staple Hill in Somerset, the eastern extremity of the Blackdown Hills. In the first part of their course the basins of the Axe and Otter have a common watershed, but from Gittisham Hill and Farway Hill, 6 miles from the coast, they are divided by a broader spread of the Blackdown Plateau which now forms a separate drainage area of triangular shape where a number of small streams, of which the Sid alone carries the title of a river, make their way to the sea between Beer Head and Otterton Point.

The 11 miles of coast between the west side of Beer Head and the east side of Otterton Point is a recessed arc or embayment which is the foreground of the view from the esplanade of Sidmouth, but invisible from the sea fronts of Seaton and Budleigh Salterton. Of this coastal foreground the Sidmouth Urban District, bounded on the east by Branscombe and on the west by Otterton Parish, comprises the $3\frac{1}{2}$ miles from Weston Mouth to Peak Hill. This recently constituted urban area comprises the three civil parishes of Sidmouth, Sidbury and Salcombe Regis. The first two fall within the watershed of the River Sid, whose source is 6 miles from the sea. The parish of Salcombe Regis shares with Sidmouth the seaward end of Sid Vale (their common boundary the River Sid) and extends farther to the east, comprising Salcombe Hill and Dunscombe Hill with their included combe of Salcombe Valley opening to the sea, and beyond this the western side of Weston Combe. The land frontier of the Sidmouth district has a length of 14 miles. Except for Weston Combe, which occupies the first mile, it follows the crest to the hills all the way round from Weston Mouth on the east to the cliff of Peak Hill on the west, a line of heights rising from 500 feet near the sea to 800 feet inland. This line of heights is smooth and unbroken except for the notch called Sidmouth Gap where, between summits of 700 feet one mile apart, the level drops to little more than 300 feet, and has thus provided the historic route connecting Sidmouth with Exeter, the capital of the county, and later with the Southern Railway system.

The area of 11,466 acres, about 18 square miles, comprised within this frontier is a plateau furrowed by combes. Most of the plateau has a thin capping of clay containing flints, beneath which is the Upper Greensand, channelled by deep combes that cut through this formation and into the underlying Red Keuper Marl of the Triassic period. The upper slopes are steep, the lower more gentle, and the trough of the valleys has a perfect V-shaped section, for there is no flat alluvial bottom except in one small portion of the lower Sid.

§ 2. During the Napoleonic Wars when Mediterranean resorts were debarred the wealthy and leisured classes went for winter sunshine to the south coast of England, those who desired more especially a mild climate seeking the south-west.

Among the places chosen was the little town of Sidmouth in East Devon, situated at the back of a shingle beach less than half a mile in length between Salcombe Hill and Peak Hill, each more than 500 feet in height.

Documents of Plantagenet times refer to the harbour of Sidmouth, but the River Sid being little more than a torrential stream, it is unlikely that there was an estuary capable of harbouring ships. The port was presumably a sea anchorage with its main shelter on the west, where the Red Sandstone reef of the Chit Rocks, still uncovered at low tide, projects far into the sea. The range of tide is small, and if the rocks have been worn down at the rate of only a quarter of an inch each year the reef would in those days have been an effective breakwater. On the eastern side of Sidmouth there was less shelter, but even here there remain some relics of off-lying rocks beyond the Red Sandstone cliff which has imposed a limit on the wanderings of the River Sid.

At the time when Sidmouth became a seaside resort there was less shelter for anchorage, but there was still a steep bank of shingle fronting the marsh or ham near the mouth of the Sid. This marsh was considerably below the level to which shingle was piled, and the permanence of the latter, so useful for launching boats, was ensured by the circumstances that during any storm which occurred when the tide was high the rampart was pushed a little further inland, not sucked away by backwash from a sea wall.[1]

The earliest account of the beginning of Sidmouth as a seaside resort is contained in the final chapters of the Rev. Edmund Butcher's *Excursion from Sidmouth to Chester*. This book appeared in the form of letters, and Butcher dates the final letter October 20th, 1803. It was written from Sidmouth, "which after all my wanderings, I am truly glad to see". He proceeds to give a description of the place. This is illustrated by a frontispiece, for which "the author is indebted to the friendly and elegant pencil of Hubert Cornish Esq.... The view is taken at low water, and from Salcombe Hill, which rises on the east side of the town. A part only of Sidmouth is included; but the beach, and the distin-

---

[1] An interesting account of the building of a small yacht at Sidmouth and its launching from the steep shingle beach is given in Robert Leslie's *A Waterbiography* (1894). Of the conditions of sea fishing from Sidmouth a realistic account is given in a work entitled *A Poor Man's House* (1908) by Stephen Reynolds.

guishing features of its coast, are sketched with fidelity and spirit."

Hubert Cornish (1757–1823), my great-great-uncle, had not been trained as an artist, landscape painting being the occupation of his leisure, and this may perhaps partly account for the fact that he did not exaggerate the slopes of hills in the way that so many professional artists of that time seemed to regard as a duty. The drawing referred to shows the low ham behind the beach as open ground. This occupies the area of what is now the eastern part of Sidmouth town. The old town, which had very little sea front, is shown running diagonally inland, approximately conforming to the foot of Peak Hill and Bulverton Hill. The population at this time was about 1250.

Butcher's description of the seaward slope of the shingle beach is an interesting record of its state before the erection of the sea wall. The beach is, he says, "...a natural rampart of pebbles, which rises in four or five successive stages from the surface of the sea at low water. With every tide the exterior parts of this shifting wall assume some different situation; are sunk either higher or lower, or are driven to the east or west, according to the strength and direction of the wind. At low water, considerable spaces of fine, hard sand are visible.... At the head of this *shingly* rampart, a broad commodious walk, which is called *The Beach*, has been constructed, and furnishes a delightful *promenade*."[1]

[1] The appearance of the beach since the erection of the sea wall and the accumulation of the shingle at the east or west end according to the direction of the wind was recorded by a careful observer, the late John Tindall (see *Trans. Devon. Assoc.* 1926, vol. LVIII).

The final chapter of *The Excursion from Sidmouth to Chester* formed the basis of the first guide book to Sidmouth, published under the title of *The Beauties of Sidmouth Displayed*. It is written in a formal style to which we are now unaccustomed, but the author displays a considerable critical faculty with reference to the aesthetics of scenery. The following is a condensed quotation from this later work:

"The immediate vicinity of Sidmouth abounds with lanes, many of them of considerable length and variety, more or less sequestered, in which the delights of solitude may be enjoyed. The *beach*, on the contrary, offers to such as are most happy in a crowd, a walk, in which, amusement and health are to be found."

As a health resort, Butcher writes:

"Sidmouth yields to none of the retreats of *Hygeia*. An air mild and salubrious, a soil uncommonly fertile, the purest water continually flowing, and a situation defended from every wind but the south, give it a pre-eminence."

When our author deals with the practice of sea bathing, which had recently become fashionable, he indulges in grandiloquence:

"In some situations, man is almost an amphibious animal and at some seasons to find ourselves plunged in the refreshing wave, and wrapped round with the liquid element, is a most delightful sensation. Health and pleasure are equally consulted in these salutary ablutions, and to many a wan countenance has the blush of the rose been restored by an occasional residence on the margin of the sea, and a frequent application of the purifying surge to the debilitated limb."

Hubert Cornish's small drawing of the sea front of Sidmouth was published in 1805. His next drawing, on a very large scale, was published separately as an engraving in 1815. It shows the way in which the housing plan of Sidmouth was changed by development as a watering place. This followed the usual custom of building as many houses as possible facing the sea and on the very front. Eastward of No. 1, Marlborough Place (where in old days the Monks of Otterton had their lodging and private chapel, of which some relics still remain), a terrace was built on the ridge of the beach, and most of the marsh behind became the eastern town, although then ill suited for habitation. On the west of the town the Fortfield Terrace was built to face the sea but at a considerable distance from the shore. The Fort Field in front of Fortfield Terrace, and other meadows between the terrace and the church, have remained unbuilt upon. These open spaces contribute greatly to the charm of Sidmouth town to-day. Being very valuable as building sites, curiosity is aroused as to the reason for their remaining thus undeveloped. It is unlikely that this was due to forethought in town-planning or to generosity on the part of landowners. A clue seems to be provided by a large-scale MS. map of Sidmouth Manor in A.D. 1789. On this the area is shown in long strips about a chain wide (a "chain" being the length of a cricket pitch). Most of these strips run westwards from the built-up area of the old town and parallel to the shore. The arrangement suggests that the strips were held by the townspeople for cultivation or pasture. Parts of several strips running parallel to the shore would have to be purchased in order to provide space for a villa facing the sea with a suitable garden attached. Moreover, experience of certain transfers of, and of failures to transfer, property in Sidmouth suggest that some of these strips may have been held "on lives", which would make their purchase for building very difficult. The most important part of the open spaces thus preserved is the Fort Field, now the well-known Sidmouth cricket ground. It gives the tower of Sidmouth Parish Church a spacious foreground, and this fine feature of the Perpendicular period is enhanced in its beauty by the background of Salcombe Hill. The natural feature is at a sufficient distance to be somewhat softened in tone, and the massive tower enhances its apparent distance, being, as French artists say, a *repoussoir*. But such effects are reciprocal, and in this particular case, according to my experience, the chief benefit resulting from the combination is the enhanced massiveness of the stately tower, erected in the culminating period of church-tower building, of stone from Dunscombe Hill.[1]

§ 3. The "Long Picture", as Hubert Cornish's panorama is commonly called, was drawn from opposite

[1] Of the Upper Greensand formation, a geological term which includes a variety of substance from light mould to building stone. The new Woolbrook Church is built of the same stone, and so we know how the tower of Sidmouth Church appeared when new.

Wallis's Marine Library, afterwards the Bedford Hotel. At the beginning of the nineteenth century it was still the custom to draw and engrave panoramic views of seaside resorts. The view which we are now considering is drawn looking landwards from the beach, that is towards the north, so that the coast from Peak Hill to Tor Bay is on the left and that from Salcombe Hill to the high ground behind Abbotsbury in Dorset on the right. The total angle comprised in the view is 240 degrees. It is of some interest to learn the method adopted for representation of two-thirds of the circumference as a rectilinear drawing. A preliminary study, 5 feet long and 10 inches high, has been preserved. This consists of two sketches each comprising about 120 degrees. These having been joined together, mounted and framed, make a picture which represents the nearly straight front of Sidmouth Beach as a sharply projecting promontory. The final drawing was evidently composed of three sketches, each 3 feet long and each comprising 80 degrees. These were reproduced in 1815 in three engraved plates, making a panorama of 9 feet. When these are framed as one there is only a very slight apparent projection of the central portion, so the problem of representation as a rectilinear drawing has been fairly well solved. The original of the final picture, in the possession of my cousin Miss Kestell-Cornish, is drawn on *six* boards (marked "Bristol paper"), but the subdivisions have so little relation to the features of the landscape that the artist evidently changed his point of sight only three times. Each sheet is $18 \times 10\frac{1}{2}$ inches, which indicates the proportion of length to height customary in landscape drawings. The vertical angle comprised in the "Long Picture" is therefore $23\frac{1}{2}$ degrees or, say, one-tenth of the horizontal angle. The circumstance that the drawing was made on six boards has facilitated its reproduction as a folded picture at the end of this volume.

Those who are interested in the natural features of the view, and in the costumes depicted, must be careful not to be misled by the second edition of the 9-foot engraving published after the death of the artist. Wallis's Marine Library had been enlarged since 1815 and Mr Wallis, the publisher of the engraving, had the central part of the plate recut so as to show his enlarged premises. The date 1840 is put upon the picture, but Chit Rock, the skerry which was washed down in the great storm of November 23rd–24th, 1824, remains, and the costumes depicted are those of the previous generation. The forms of the fishing boats shown in the picture are the same as we now see hauled up on the broad belt of shingle where the Sid reaches the sea. They are clinker built, and each plank overlapping that below displays the same stream-line curvature that Nature has adopted in the forms of aquatic animals.

Among the architectural details in both the first and later editions of the panorama are the stumpy pinnacles of the church tower, which are those that I first remember. They were replaced in 1873 by the tall, elegant pinnacles which now cap the tower so suitably. These are conformable with the Perpendicular style of the tower, the stumpy pinnacles existing in 1815 being then a recent replacement of earlier work.[1]

One important character of the scene depicted in the original drawing remains to the present day, the absence of houses on the actual sea front beyond Sid Vale. Such immunity from building is very rare in the neighbourhood of our south-coast watering places, and it is of great importance to the amenity of Sidmouth to secure its continuance.

Panoramic views, a favourite form of picture with Chinese artists, are no part of the ordinary output of English artists to-day. This to some extent helps to explain the circumstance that Sidmouth, with a wide coastal prospect which delights the layman, is not a chosen haunt of artists, as is, for instance, St Ives in Cornwall. There the eye is attracted by numerous views of from two to three spans breadth. The angle comprised by the outstretched hand at arm's length is about 18 degrees, three spans making 54 degrees. Sixty degrees, the angle of an equilateral triangle, is a wide angle for a landscape picture. As the panoramic view, which necessitates a roving glance, does not readily lend itself to the painter's art, it is desirable to encourage and confirm the layman in his instinctive appreciation of these broad prospects. The essential truth is that the eye is capable of receiving truly pictorial effects of which some do not lend themselves to treatment by pencil or brush.

[1] These "little absurdities" are discussed in an instructive pamphlet, *History of the Restoration of Sidmouth Parish Church* (1860), by P. O. Hutchinson.

§ 4. When terraces were erected on the sea front, or perhaps a little earlier, villas were built in neighbouring fields of Sidmouth Parish. Many of them are depicted in Edmund Butcher's illustrated volume on *Sidmouth Scenery, or Views of the Principal Cottages and Residences of the Nobility and Gentry of that Admired Watering Place*.

Most of these houses still remain, dotted about among the more numerous villas which were built in late Victorian times and in the twentieth century. The "cottages" of the gentry built in the latter part of the eighteenth century and beginning of the nineteenth, although not actually beautiful, have a certain attraction as illustrating a phase in the aesthetics of scenery. The poet Coleridge (who knew the neighbourhood in this period, having been born at Ottery St Mary, A.D. 1773) describing *The Devil's Thoughts*, writes that

"He saw a cottage with a double coach-house,
A cottage of gentility!
And the Devil did grin for his darling sin
Is the pride that apes humility."

The poet's sarcasm is unjustified. It was not from false modesty that people of ample means built their country residences in cottage style but in recognition of the charm and benefit of arcadian life. It helped them to forget the turmoil of the town. The spirit was much the same as that which has led many well-to-do people in the twentieth century to recondition country cottages for their own habitation.

The cottages of the gentry erected when Sidmouth

first became a pleasure resort were mostly thatched, following the Devonshire fashion. In other respects they embodied the style of "Strawberry Hill Gothic", especially in their pointed windows. A good example of the neo-Gothic window is provided by the Royal Glen, as it is now called, the house in which Princess Victoria, afterwards our Queen, stayed in her early childhood.

Another interesting feature is the up-curving roof of the verandahs and porticoes, an example of the chinoiserie which came into fashion after intercourse with China had been promoted by the East India Company.

The surface of the ground in the inland parts of Sidmouth Parish, where the earlier and later villas have been built, is largely in mamilliform undulation, an alternation of cups and bosses. This is doubtless due wholly or in part to the outcrop of New Red Sandstone, which weathers in this manner. The eminences provide situations which give many of the houses an extremely attractive southern outlook on the sea, with Peak Hill and Salcombe Hill framing the picture. Perhaps the finest of all these sites is that of Sidmouth Manor House. This house is brickwork of late Victorian times, dignified in design but lacking the attractiveness of colour and relief which ancient, and some of the most modern, brickwork shows. The grounds sloping steeply towards the sea remind one, especially on a fine winter day, of those surrounding some choice residence upon the bold shore of the French Riviera.

Many of the Sidmouth villas are more than a hundred years old, and the growth of the trees planted in their gardens has greatly altered the outlook from these houses.

The trees, rooted in fertile Red Marl, grew rapidly, and many of them are now of great size, shutting out a view which was originally the chief attraction of the site. But Englishmen are tree lovers. Of all features of scenery it is the forms of the forest that we, as a nation, specially appreciate. The most remarkable trees in the old-world gardens of Sidmouth are the pines. These have attained great girth as well as height, and the branches spread so widely that the tree is much more umbrageous than those of "sand and pine" districts such as Bournemouth and Camberley. Their growth resembles that of pine trees in Japan, of which the special attraction to the Japanese is the appearance of protective shelter imparted by the spreading branch.

It has long been the general feeling in England that to cut down a fine tree is an act of vandalism. The park-like appearance of our countryside, which largely results from this taboo, is an aspect of our scenery which specially attracts the visitor from the Continent. But the Englishman's love of trees is lacking in discrimination. They are sometimes planted in the wrong place, and not sufficient courage is shown in the thinning out of their growing screen. Not only in the gardens, but also in the countryside of Sidmouth, trees have been allowed to shut out views which should have been kept open. Salter's Cross, between Mutters Moor and Bulverton Hill, a mile and a half from the sea front, used to be a favourite viewpoint for the sake of the wide prospect of East Devon, but a fir plantation on the slope west of the brow now hides the view. However, the way thereto up Bickwell Vale is still a charming walk in the shelter of the flowery banks characteristic of Devonshire lanes.

As a matter of Town and Country Planning it is desirable that the skyline of the plateau which surrounds Sid Vale should be kept as free from trees as possible, for objects of such well-known size tend to scale down the apparent height of the hills.

Returning to the subject of villas—it should be noted that the fertile soil and mild climate make the flower garden develop its beauty with surprising rapidity, delighting those who settle in Sidmouth for the leisure of their later years. In the gardens of Sidmouth the song of the birds continues into the winter months, an unexpected delight to those who come from the colder parts of England and an unfailing solace to the residents.

Here I may remind the reader that the picturesque impression of scenery is greatly enhanced by the pleasures of sound. The enjoyment of the late song of birds in the Sidmouth countryside was recorded more than a century since in the stanzas *To the Red-Breast* by my great-uncle, the Rev. George James Cornish (1794-1849), which appear as a supplement to the poem for the Twenty-first Sunday after Trinity (an autumnal date) in Keble's *Christian Year*:

"Unheard in summer's glaring ray,
Pour forth thy notes, sweet singer,
Wooing the stillness of the autumn day:
Bid it a moment linger,
Nor fly
Too soon from winter's scowling eye."

The architectural background of the villadom of Sidmouth Parish is provided by farmhouses and by the cottages of agricultural labourers. The cottages are of cob and thatch, cob being rammed earth. The thatch is well laid, and of straw uncrushed by thrashing. Many of the farmhouses show the distinctive feature of the outside stone chimney-stack, which, regarded from the decorative point of view, is a tower, and thus dignifies the building. As a novel feature in domestic architecture and one to be proud of, the date of erection, generally seventeenth century, is cut in bold figures.

CHAPTER II

SIDMOUTH PARISH (*concluded*)

§ 5. The Sidmouth esplanade, now protected by a sea wall, faces 20 degrees east of south, that is to say about south-south-east. The stretch of the sea between the heights of the Dorset mainland near Abbotsbury and the promontory called Downend Point or Scabbacombe Head, which marks the entrance to the Dartmouth estuary, comprises an angle of nearly 120 degrees. From time to time, and as a sign of rain according to local lore, the distant wedge of Portland is seen rising out of the sea as an island, for the Chesil Beach which links it to the land is not visible. The highest point in Portland (distant 35 statute miles from the Sidmouth esplanade, as I measure on the chart) is 459 feet above mean sea level. The distance of its horizon, allowing for average refraction, is 29 statute miles. Taking the eye-height of a spectator on the Sidmouth esplanade as 25 feet above mean sea level, the distance of the sea horizon under the same conditions is 6·7 statute miles. These figures add up to 35·7 statute miles, which shows that the highest point of Portland can be seen from the Sidmouth esplanade under average conditions of refraction when the air happens to be perfectly clear. When, however, Portland is clearly visible, a fairly long piece of the wedge-shaped promontory appears, which, as far as my opportunities of observation go, indicates that we can sometimes see all that stands above the level of about 400 feet. Conditions of refraction near the horizon are so variable that the compilers of mathematical tables are chary of giving maximum figures.

In some of the earlier accounts of Sidmouth the limit of the western view is stated to be Start Point, but in fact this is hidden by the promontory which has the alternative names of Downend Point and Scabbacombe Head. In many of the more recent accounts of Sidmouth the western limit of the coastal view is stated to be Berry Head on the farther side of Tor Bay. This mistake may be due to the circumstances that at night the flash of the lighthouse on Berry Head comes from a point very near the daylight limit of the coast. No one, however, who has visited Berry Head and noted the abrupt termination of its vertical cliff would identify this promontory with the sloping point of Downend which extends a little farther. The name Downend Point is used on the Admiralty charts and in the Admiralty *Channel Pilot* (Part 1). In maps based on Ordnance Survey it is named Scabbacombe Head, the name "Down End" being given to the high ground a little farther inland. In *The Handbook* of the Norman Lockyer Observatory on Salcombe Hill, Scabbacombe Head is the name given to the western termination of the coastal view.

The opening of the semicircular bay between Portland and Scabbacombe Head is 48 statute miles, the beach from Portland to Sidmouth 43, from Scabbacombe Head to Sidmouth 40 statute miles. Whether looking east or west from the Sidmouth esplanade the view of the coast consists of a bold foreground showing local colouring and a distant background in atmospheric tint.

Eastwards the first feature is the bold outline of Salcombe Hill. The cliff is the highest on the south coast of Devon, 540 feet above mean sea level. In North Devon there are loftier hills sloping steeply to the sea of which only the lower part shows a rock escarpment. Mr E. A. Newell Arber, in *The Coast Scenery of North Devon*, describes these steep slopes covered with vegetation as "Hog's Back Cliffs". It seems impossible to make any definite rule as to a limit of steepness which will enable one to discriminate between a "Hog's Back Cliff" and other hill slopes descending to the sea, and I consider it preferable to restrict the term "cliff" to a rocky escarpment as in the *New English Dictionary*, our standard authority.[1]

The summit of the plateau of Salcombe Hill slopes very gently, almost imperceptibly, seawards, but the skyline as seen from the esplanade rises in the field of view to the very verge of the cliff, and so the lines of sight lead from either hand up to the point of most interest. Thus "decoration and significance", as the late Sir Charles Holmes used to express it, combine in the enhancement of effect.[2] Moreover, the sloping upper

---

[1] *N.E.D.* "CLIFF. I. A perpendicular or steep face of rock of considerable height, usually implying that the strata are broken and exposed in section; an escarpment.

b. *esp.* (in modern use). A perpendicular face of rock on the sea shore, or (less usually) overhanging a lake or river."

[2] *Notes on the Science of Picture Making*, by C. J. Holmes.

portion of the cliff is a *batter* surmounting the wall of rock which rises steeply from the shore, and the jutting elbow thus produced imparts the appearance of a defensive rampart of the land.

Beyond Salcombe Hill the cliff rampart extends to the chalk promontory of Beer Head, $6\frac{1}{2}$ statute miles in a direct line from the centre of the Sidmouth esplanade. This line of lofty cliff is furrowed by steep-sided combes, of which the seaward opening or "Mouth" is visible in the sideway view, these notches dividing the cliff rampart into five blocks.

Looking west from the Sidmouth esplanade, the coastal foreground is a triad of forms, Peak Hill, High Peak, and the rocky promontory of Otterton. This promontory is only $4\frac{1}{2}$ statute miles distant in a direct line from the centre of the esplanade, but, as the direction of the coast line changes abruptly with the outcrop of the hard Red Sandstone at the base of High Peak, the triad of forms on the west subtends a greater angle from the esplanade than the pentad of forms on the east, a circumstance which helps to maintain the scenic balance of the coastal panorama.

The cliff of Peak Hill is 526 feet above mean sea level, which is only 14 feet lower than that of Salcombe Hill, yet its appearance from the esplanade is much less impressive. The cliffs differ little in colour or in steepness, but the crest line of Peak Hill as seen from the esplanade slopes gently down to the edge of the cliff, so that the cliff summit is not the culmination of ascending lines and does not present the appearance of a "peak" in the sense in which that word is usually employed in reference to hill forms.

The contour lines are so spaced on the eastern and western sides of both Peak Hill and Salcombe Hill that each of them when seen from the west appears "peaked" at the cliff, whereas viewed from the east the ridge in each case slopes down slightly towards the cliff.

The appearance of High Peak from the esplanade is truly pyramidal, and as the cliff does not continue the line of the esplanade but turns about 60 degrees to the left hand, this boldly sculptured crag of dark red rock is much better seen than the face of any other cliff in the foreground of the coastal panorama. Facing the promontory stands Picket Rock, a castle in the sea where the ravens build their nests. Other skerries of New Red Sandstone lying close to the picturesque shore of Ladram Bay are hidden from view by High Peak.

The scenery of a place means, properly speaking, its aspect and outlook. So far I have spoken of the prospect or outlook of the Sidmouth coast looking to right and left from the esplanade. No position on the line of cliffs to the east affords an open view of the Sidmouth coast, which is seen greatly foreshortened. It is otherwise with the jutting promontory of Otterton on the west. This is the one place on the neighbouring shore which provides an open panoramic view of the Sidmouth coast. In addition to the cliffs visible from the sea front of Sidmouth the recessed portion of Dunscombe Cliff, which forms the eastern limit of the sea front of Salcombe Regis, can be seen—an important part of the Sidmouth scenery which will be dealt with in some detail in a later chapter.

§ 6. The Sidmouth esplanade, as already mentioned, faces 20 degrees east of south, that is to say nearly south-south-east. All the year round the sun sets behind the coast. On the east the mainland of Dorset only extends 10 degrees south of east,[1] and so the sun rises from the sea during nearly five months of the year, from the second week of October to the first week of March. Then from the sea front, when the sky is clear, we can observe the emergence of the sun's upper limb above the horizon of the sea, the most spectacular event in all the round of Nature's day. A lane of rosy light is thrown across the water from the horizon to the breaking wave, and the cliffs of High Peak and Otterton are flushed by the glow of the rising sun.

At certain times of the year the moon, low down, spreads a broad band of silver across the sea against which the sails of the fishing boats are seen in dark silhouette.

In the nocturnal view, Sidmouth shares the advantage common to most south-coast watering places of an esplanade which provides an observatory from which to watch the culmination of the constellations above the broad, level horizon of the sea. The finest effect of all is when, during winter nights, the form of Orion stands upright on the meridian with the attendant constellations which combine in one great pictorial pattern, from the Pleiades which flee before the Hunter to Procyon and Sirius, the hounds that follow after.

[1] The Isle of Portland, somewhat farther to the south, is seldom visible.

Few perhaps but those who have seen the constellations in various latitudes realise the advantage which we enjoy from the aspect in which Orion is displayed to us, particularly on our southern shore. Here the constellation reaches 45 degrees, about as high as is consistent with an upright as distinguished from an overhanging effect. Seen at the equator this constellation rises on its side, passes practically out of sight overhead when on the meridian, and sets on its side; its characteristic symmetry never suitably displayed. The flash of Sirius moreover, that splendid culmination of the sparkle of the stars, is less striking in more southern latitudes where it stands higher in the sky. There Canopus takes the lead among the flashing stars, but with less luminosity.

No description of the scenery of Sidmouth would be complete without some examination of the various characteristics of the waves which reach us here according to the weather, the tides, the direction, and the slope of the shore. Waves travelling before the westerly wind, and the swell after gales in the Atlantic, are reduced in height as they swing round Otterton Point and roll in towards Sidmouth beach. They do not exceed the height of the waves travelling down Channel from the east as greatly as would be the case if the Sidmouth beach were equally open on both sides. The interval of time between the waves from the Atlantic is, however, notably greater than that of the waves raised by easterly gales, and so the rhythm of the surge with a westerly swell is slower and more solemn.

The height of the breakers increases as the tide rises

from the gentle slope of the lower to the steeper slope of the upper part of the beach. Then the cusp of the wave is seen in its greatest perfection, marbled with foam on the smooth surface of its hollow front, ribbed and corded on the curling crest which trembles when about to fall.

The sound of the sea changes also as the tide rises, a murmur at low water, then a rhythmic beat, later, at high water, less musical, for the shingle rasps and rattles in the backwash of the wave.

Against the western end of the sea wall, and the projecting cliff of New Red Sandstone beyond Clifton Terrace, the storm waves burst and fling the spray high into the air.

Between the reef called Chit Rocks, bare at low water, and the projecting cliff of High Peak, the shore is shallower than elsewhere in Sidmouth Bay, and here at low water of spring tides the waves come sliding in over smooth and almost level sand, not as curling breakers but as long, low foaming ridges, which do not undulate but travel independently, rank behind rank, thinning as they come, till at last each leaves a line of froth upon the beach. This charming effect can be seen from Connaught Gardens, the delightful place of rest which crowns the projecting bluff of Red Sandstone rock overlooking Chit Reef.

The condensation of atmospheric moisture by the cliffs produces effects which add greatly to the scenic charm of Sidmouth. During fine weather, mist gathering from the sea in the early hours of the day sometimes spreads a white cloak across the valley. As the sun gathers strength the hill tops are seen, aetherial, through the thinning mist, and then appears a sky of the pure blue characteristic of coasts near to the dust-free Atlantic ocean.

During wild weather the inland scenery may take on a dreary aspect, but the gathering of cloud against the cliffs has a gloomy splendour which appeals so strongly to the imagination that it casts no depression upon the observer but raises him to the appreciation of the poetry of Nature in her sterner moods. It is against the long rampart of the eastern cliffs of Salcombe Regis and Branscombe that the brooding clouds are most striking. Wreathing over the crest of Salcombe Cliff the cloud caps the hill with that stream-line form which escaped attention fifty years ago but leaps to the eye of the present generation, which learnt to appreciate its significance as it successively appeared in the aeroplane and the racing car. The trace of the stream-line remains in the over-arching boughs of a fence by the cliff on the way up Salcombe Hill. The growth of the fence is limited to the eddy-space beneath the up-sweeping currents of the on-shore wind.

There is another and more elementary effect of the capping of the hill by a cloud. It establishes a marked distinction between the regional scenery of lowlands and of hilly country. This was strongly impressed upon me in childhood's day when I used to come from my father's vicarage in flat Suffolk to spend the summer holidays at Salcombe House, my grandfather's home in the valley of the Sid. It was largely because the hills touched the clouds that I was thrilled by the scenery of Sidmouth.

§ 7. There was another aspect of Salcombe Hill which impressed my imagination when a boy, its *solidity* when viewed from the sea front. In Suffolk the landscape was a mere surface, here the cliff front showed the solid geometry of the world. I felt that solid hill to be far finer than a mere surface, and this impression of childhood is confirmed by experience. Cliff scenery has the same superiority over an inland landscape as sculpture in the round compared with a bas-relief. It is also preeminent in its contribution to the scientific understanding of the origin and nature of the ground we tread. The cliffs from Lyme Regis to Sidmouth so clearly show the structure of the plateau which they terminate that they led the pioneers of geology to write certain chapters in the physical history of the world. As we look along the line of cliffs from Branscombe Mouth westwards to High Peak the different colour of the strata shows the succession of the rocks, their upward tilt towards the west is evident, as well as the thinning of the Greensand and the appearance of the New Red Sandstone below. The change from the Red Marl to the Greensand above, so conspicuous on the cliffs owing to the change of colour, is equally evident on the hillsides from the steepening of the slope. Indeed there are few districts where the geological map and the map of relief so closely correspond.

In 1822 the Rev. William Buckland, Professor of

Geology in the University of Oxford, read a paper before the Geological Society, "On the Excavation of Valleys by diluvial Action illustrated by a succession of Valleys which intersect the South Coast of Dorset and Devon".[1] He writes:

"I beg...to present to the Society two geological views of that coast drawn at my request some years since by Hubert Cornish, Esq., which will tend... further to illustrate the description given of it in a preceding paper by Mr De la Beche."

The author had visited Salcombe Regis in 1819, whence he went to Dunscombe Cliffs to prospect the scene. In the drawings to which he refers the cliffs are coloured so as to show the strata. That taken from the top of High Peak includes the cliffs from Peak Hill to Beer Head. It is a remarkably accurate panorama, clearly demonstrating the geological structure of the cliffs, whilst its perspective imparts a realistic effect which is lacking in a sectional diagram.

The special purpose of the paper was to emphasize the author's conclusion that the combes debouching on the sea could not have been produced by gentle erosion such as occurs at the present time, but were formed by "diluvial", or as we now call it, torrential action. That is as far as Dean Buckland's opinion was expressed in the original paper, but in the volume entitled *Reliquiae Diluvianae* which he published in 1823, in which Hubert Cornish's drawing again appears, the substance of the paper is repeated as part of the geological evidence of a

[1] *Trans. Geol. Soc.* Ser. 2, vol. I, pp. 95-102 (read April 19th, 1822).

Universal Deluge, that is to say of the Biblical narrative of the Flood. The idea that the Babylonian flood drowned the whole world has long been abandoned, and the student of mid-Victorian times would have dismissed the opinion of Dean Buckland as an entirely mistaken idea engendered by a foregone conclusion. Later investigation, however, points to the fact that in the dissection of the Greensand plateau a period of torrential erosion did occur, and at a date geologically so recent that it did not differ greatly in time from that assigned to Noah in Archbishop Usher's chronology. The late Mr Clement Reid of the Geological Survey stated definitely that the last change of sea level on our southern coast (which would be followed by torrential erosion) "only took place in Neolithic times, and probably only some three thousand years ago".[1] The late Mr A. W. Clayden, writing some years afterwards, states that the glacial period in England ended "not very many thousand years ago".[2]

On the 1-inch Ordnance map the 5-fathom line is marked, and this is significant in relation to the Sidmouth embayment, for it approaches to within about a quarter of a mile of the compact chalk of Beer Head and the firm New Red Sandstone of Otterton Promontory at Danger Point, but is more than twice this distance from the soft Red Marl cliffs of Salcombe Hill and Peak Hill. It is noteworthy that the great cliffs of Salcombe Regis and

---

[1] *Geographical Journal*. Discussion of a paper "On Sea, Beaches and Sandbanks", by Vaughan Cornish (read March 16th, 1898).
[2] *The History of Devonshire Scenery* (1906), pp. 180–81.

the parish of Branscombe do not form promontories, the curve of the coast being continuous whether the height be great or, as at the "Mouths", almost negligible.

The explanation is that the recession of these cliffs is not entirely due to the scour of the tides and the battering of the waves, but largely to weathering and the percolation of inland waters. That the waste by weathering continues unchecked is due to lack of binding vegetation, and this in turn is caused by the blast of winds laden with the salt of the sea. The losses of a cliff are always lamented, but consolation should be found in the fact that the continual renewal of surface maintains the freshness and richness of colour which is so great an asset in cliff scenery. The red cliffs of the Sidmouth coast do so much to redeem the dullness of the English winter that they provide some compensation for the sunshine of the Mediterranean shore. Every characteristic of scenery, however, has the defects of its qualities, and I have known glaring days in August when one would have preferred the cool colour of the chalk. This defect is however only rarely felt, and taking the whole round of the seasons the red cliffs and red soil of East Devon are a welcome addition to the colour scheme of the coast and countryside.

At one point, I have been able to make a measurement of the rate of coast erosion by comparison of the Ordnance Survey map of the 1933 edition with the large-scale map of Sidmouth Manor dated 1789. This shows a recession of the cliff edge of the lower slopes of Peak

Hill (where the Red Marl succeeds the New Red Sandstone of "Jacob's Ladder") amounting to 62 feet in 144 years, that is to say, at the rate of 43 feet in a century. Within living memory the falls of cliff from Peak Hill have been greater than from Salcombe Hill, and it seems likely that this may be connected with the wearing down during preceding centuries of the reef of New Red Sandstone known as Chit Rocks which has exposed the neighbouring cliff of Peak Hill increasingly to the attack of waves from the east.

§ 8. The cliff summit of Peak Hill is an extremely attractive viewpoint. The broad expanse of sea, the varied outline of coast, the bold foreground of the High Peak crags, are seen from the entirely appropriate surroundings of an open hill top of gorse and heather with winding paths of rough grass between. I have known it thus for seventy years, but in these iconoclastic days no such shrine of scenery is safe unless its preservation is guaranteed by an agreement with a Local Authority or a covenant with the National Trust. Only half the area is within the Sidmouth Urban District but it is all in one ownership and suitable arrangements to guarantee the preservation of its amenity for all time should not be difficult.

The modern outlook on scenery is largely that of the motorist, whose glance, though cursory, is more comprehensive of England's countryside than that of his predecessors, and it is therefore relevant to note that whereas the motor road from Sidmouth over Salcombe Hill and Dunscombe Hill runs far inland that crossing

Peak Hill runs much nearer to the cliff and affords a prospect of the great bay from Portland to Downend Point. It also provides both a view of Sid Vale and an outlook on the more ancient geological district which lies to the west of the Greensand plateau.

In order to understand the geological character of the general view of Sid Vale it is necessary to trace the outcrop of the Upper Greensand from east to west in Southern England. It is almost everywhere on the flank of a hill which is capped with chalk. The most notable exception is the Greensand plateau already referred to which extends from the ridge called Blackdown Hills overlooking the plain of Taunton in Somerset, down to the stretch of coast which extends from Lyme Regis to Sidmouth. This is the only considerable area in England where the Greensand crowns the hill, and on the coast it is only from Seaton to Sidmouth that we have the particular form of hill produced by the sculpture of Greensand above and Red Keuper Marl below. This form, with its double slope, steep above and gentle below, has been well shown in the photograph of Buckton Hill in Sid Vale which has been chosen as an example of the "Characteristic Greensand capped Hill" in Plate XI of the *Regional Geology of South West England*, published by the Department of Scientific and Industrial Research (1935).[1]

Let us now consider the western view from Peak Hill. The whole breadth of the valley of the Otter is spread out before us. Beyond its western watershed the farther

[1] Unfortunately the hill is there misnamed Trow Hill.

boundary of the Exe Valley can be seen, all this foreground and middle distance being the characteristic undulating country of "the New Red Rocks". The weathering of these has produced the rich red soil which is so distinguishing a feature of East Devon.

Beyond the basin of the Exe the heights of the southern promontory of Devon can be seen on the left and the granite boss of Dartmoor on the right. Still further to the right Pinn Hill near by hides Exmoor, which however comes into sight in the prospect from Beacon Hill, to be presently described. Exmoor, like the southern promontory, is composed of hard rock, as are all the formations of Carboniferous and pre-Carboniferous age as well as eruptive intrusions of later date such as the granite of Dartmoor. The line which separates hard from soft rocks is a frontier of great importance in the scenery of England. It was chosen as a line of demarcation by H. J. Mackinder in a map entitled "The Older and Newer Rocks of Britain", the latter being the post-Carboniferous.[1] The eastern members of the New Red Rocks, the Triassic, are undoubtedly of the Secondary Period. The western, or Permian, are classified by some geologists as Secondary, by others as Palæozoic. If the latter classification be adopted it is certainly preferable to divide the scenery of Devon by the post-Carboniferous rather than by the post-Palæozoic line, for the landscape of Liassic and Permian sandstone is essentially one, whereas the change from Permian to Carboniferous and pre-Carboniferous is very marked.

[1] *Britain and the British Seas* (2nd edition, 1915), Fig. 34, p. 64.

Another source of confusion in the scenic division of Devonshire arises not from chronology but terminology. Both Exmoor and the southern promontory which terminates in Start Point are usually described on geological maps as "Devonian or Old Red Sandstone". If therefore the tourist has been studying the geology he will expect to find the scenery of these parts characterized by rocks of red colour. In Herefordshire, indeed, the rocks of the Devonian period, there deposited under estuarine conditions, are in fact sandstones of a red colour, but in Devonshire "Old Red Sandstone" is a misleading expression, for there the slates, grits and sandstones of the period, deposited under marine conditions, are not of a red colour.

On the skyline, beyond the undulating landscape of East Devon, one feature arrests attention, the triad of Haytor, Saddle Tor and Rippon Tor where the granite of Dartmoor stands out in rugged peaks at a distance of 23 miles. Their height, less than 1600 feet, is below the limit which we are now accustomed to associate with the term "mountain", but the appearance of this group of rocky peaks crowning the treeless slopes of the granite moor is of a strikingly mountainous character. The impression of sublimity which I received when a boy remains undiminished now that I have seen the Alps and the Rocky Mountains. There is indeed nothing in the view to provide a scale of measurement which would reduce their apparent size, and an uninformed visitor from mountainous lands would probably suppose their height to be much greater than it is. In this connection

I recall a conversation with the late Mr Douglas Freshfield, of mountaineering fame, who told me that his Alpine guide, when viewing Arthur's Seat from Edinburgh, estimated the time required for its ascent at three hours, although the height of this rocky eminence is only 822 feet.

Turning from the physical to the historic aspect of the scene from Peak Hill, the first question which occurs is the origin of the road which brings the motorist within sight of the cliffs. Nowadays we think of the road as a connection between the watering places of Sidmouth and Budleigh Salterton which happens by chance to pass through the village of Otterton. In pre-Reformation days however the importance of Otterton as compared with these watering places was very different from what it is now. William the Conqueror had bestowed the Manor of Sidmouth upon the Abbot of St Michael's Mount in Normandy, whose deputy was the Prior of Otterton, and from Otterton Priory the monks came by the road over Peak Hill to their Rest House in Sidmouth for the management of manorial business.[1]

§ 9. Following the frontier of the Sidmouth district northwards from Peak Hill we come to Sidmouth Gap, the deep notch between Bulverton Hill on the south-west and the ridge of Beacon and Core Hills on the

[1] We are indebted to the late Peter Orlando Hutchinson (1810–1897) for his tireless investigation of this and other matters relating to the history of Sidmouth. He twice visited St Michael's Mount in Normandy for the purpose of copying documents. Hutchinson was unable to publish his *History of Sidmouth* which remains in MS. form in the public library of Exeter. Much of it is however now being printed in *The Sid Vale Monthly*, published by the Sid Vale Press, Old Fore Street, Sidmouth.

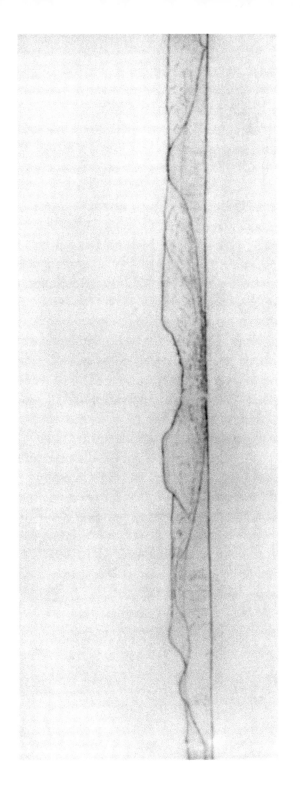

SIDMOUTH GAP FROM THE SEA

north-east through which the railway comes. The slope to the sea being somewhat steep the railway was not extended to the town, the station being placed nearly a mile from the front. The Station Road, sunk deeply in the Red Sandstone and well shaded by trees, provides an attractive approach to the town.

The Gap is itself a marked watershed. From the hamlet of Bowd, where the road from Sidmouth branches out for Exeter and Ottery, the Woolbrook drains eastwards to the Sid, and on the west a number of brooks cutting deep gulleys in the New Red Sandstone unite in a stream which joins the Otter, flowing through the charming Harpford Wood where Sidmouth people gather the wild flowers of spring. Here the rich yellow of the primrose and the size of its bloom surprise those who come from places of colder climate and less fertile soil.

The New Red Sandstone reaches the surface on the east side of the watershed, and the idea inevitably occurs that the intrusion of this rock may have some connection with the formation of the only break in the western rampart of Sid Vale.

The Woolbrook joins the Sid at Livonia Ford. The boundary of Sidmouth Parish follows the crest line of Beacon and Core Hills and thence drops down in almost the same direction to this junction of the streams.

A striking view of the Gap is obtained from the much frequented footpath which runs from a point on Salcombe Hill Road opposite to the Norman Lockyer Observatory to the Salcombe Hill cliffs. The bold incision, although not as impressive as an eminence of equal dimension, is nevertheless a striking scenic feature, and the Gap viewed from the Salcombe side of Sid Vale has an attraction of its own which an eminence cannot offer, for its steep sides are the frame of a landscape picture. Moreover, this is the only picture of the neighbouring country of the Otter which can be seen from the eastern side of Sid Vale.

Of all aspects of Sidmouth Gap the most striking is that obtained from several miles out at sea. In the summer of 1896 I chartered a small yacht at Poole in Dorset for a month's cruise along the Channel coast and when passing Sidmouth at a distance of about 9 nautical miles observed that the cliffs of Salcombe and Peak Hills were not conspicuous, being backed by the level line of the inland plateau, and that the only conspicuous feature in the skyline of the land was Sidmouth Gap, as is shown in the accompanying illustration reproduced from the sketch which I made at the time. The Gap appears much deeper than it seems when viewed from Sid Vale, for at this distance from the coast the height of the Gap above sea level and the height of the hill above the Gap are seen in true proportion.

Thirty-seven years later, that is to say in 1933, when staying in the Exmoor district, friends took me by motor to the top of Winsford Hill. The expedition had not been planned, and I had no knowledge of the features other than Dartmoor which would appear on the horizon. I supposed, in my ignorance, that the southern skyline was somewhere in the middle of the county, but a tiny notch in this skyline attracted my attention by a peculiarity of form, that of a double slope with the steeper half above, which, as I have already mentioned, is characteristic of combes cut where Greensand caps Red Keuper Marl. On consulting the map which I had with me I found that I was looking at Sidmouth Gap, 30 miles away, a distance at which the 400-foot incision subtends about 9 minutes of the arc, less than a third of the moon's diameter.

Beacon Hill, the eminence north of Sidmouth Gap and on the border of Sidmouth Parish, commands a complete view of the country to the west. The treeless heights of Dartmoor and Exmoor, which lie respectively south-of-west and west-of-north, together occupy about half of the land view beyond the Sidmouth portion of the Blackdown Plateau and its northern continuation. The extent of the prospect, the boldness and variety of form, richness of colouring and delicate gradation of atmospheric tone produce an impression upon the sense of sight which exalts the mind above the mood of common things. Moreover, the imagination is stimulated by the story which the scene unfolds. The distant heights of the southern promontory, Dartmoor, Mid-Devon and Exmoor, are the shoreline of an ancient land which was washed by the waves of the Liassic Sea whose waters covered all that is now East Devon.[1]

[1] The map in A. W. Clayden's *History of Devonshire Scenery* entitled "Ideal Restoration of Liassic Geography" shows West Devon and Cornwall, Wales and the Pennines as shores of the Liassic Sea. In H. J. Mackinder's *Britain and the British Seas* a similar map of land and sea is described as relating to the "Jurassic epoch", which comprises the Liassic and Oolitic periods.

CHAPTER III

SIDBURY

§ 10. Sidbury, the largest of the three civil parishes which constitute the Sidmouth Urban District, is entirely inland. On its southern side is included the village of Sidford situated on the reputed Roman road which runs east from Sidmouth Gap through Salcombe Regis and on to Colyford where it crosses the Coly and the Axe on the way to the ancient port of Lyme Regis. The old pack-horse bridge at Sidford has fortunately been preserved. It crosses the Sid just below the junction with the Snodbrook which drains Paccombe-Chelson and Harcombe-Sweetcombe valleys on the east.

Through Sidford also runs the old highway from Sidmouth to the north which reaches the village of Sidbury a mile farther inland, thence runs up Sid Vale, over the crest of the plateau and down to Honiton, where it joins the important Roman road connecting Ilchester with Exeter.

The status of Sidbury as an important village centre of an agricultural district is indicated by the dignity of its ancient church. Its steeple is the architectural focus of the inland half of Sid Vale, and the spirelet which caps the tower makes this steeple a unique feature in the Sidmouth district.

The cultivated fields, the clustered cottages of the

village, the mediaeval church and the Court Hall, which was once the Manor House, are all related to that aspect of rural England which we owe to the Saxons, who cleared the tangled brushwood of the valleys and ploughed the stubborn glebe. Before their time the open land of the plateau surrounding Sid Vale was the human habitation. The chief monument of man's activity in those days is the Celtic earthwork called Sidbury Castle which crowns a detached outlier of the ridge on the west of Sidbury village. The flanks of this ridge, which forms the true right-hand watershed of the Sid, differ remarkably on the western and eastern sides, the former, a long, uninterrupted slope; the latter furrowed by a dozen goyles where little streams, resulting from the south-eastward dip of the strata, break out at the foot of the Greensand and are tributary to the Sid. The first promontory on the north of Sidmouth Gap, Core Hill, is connected almost on the level with the main ridge which it joins at Beacon Hill, and therefore would not have lent itself to defence on the western side. The next eminence, which is crowned by Sidbury Castle, has been isolated by natural erosion, which has removed the whole thickness of the Greensand on the west, so that it has a steep approach on all sides. It has also a unique advantage of outlook, for no eminence interferes with the inland view, and, projecting farther to the east than Core Hill, it commands an unobstructed view of the considerable stretch of sea which is framed between the slopes of Peak Hill and Salcombe Hill.

Unlike the British encampments of the chalk country,

Sidbury Castle no longer stands conspicuously upon a bare down but is completely hidden by a thick growth of trees on its ramparts. The encampment, once the outstanding feature of historic interest for the people of Sidmouth and Sidbury, having passed completely out of sight, nothing remains in the valley to stir the mind with memories of ancient Britain. Cut the trees down and keep the ramparts clear of bracken, and the fortress crowning the lofty hill would once more be conspicuous! It is singularly appropriate that it would be a specially noticeable feature in the view from the mansion house of Sidbury Manor, the estate in which the camp is situated. In the south-west of England there are many British camps thus masked by growth of trees. If Sidbury Castle were cleared the people of Sid Vale might well be proud of giving the lead in a new development of scenic preservation.

§ 11. In the village just above Sidbury Church the River Sid is joined on the west side by a tributary of which the ultimate source is in Lincombe Goyle, a mile and a half distant on the north-west. Above its junction with the Sid this tributary is joined on its true right bank by a stream which drains the depression between Sidbury Castle Hill and Bald Hill. The next tributary, also on the right bank, drains the depression between Bald Hill and the projecting ridge which is covered by the wood called South Lincombe Plantation. The third and last tributary, also on the right bank, flows down Beacon Goyle (not connected with Beacon Hill). Above the junction of this last tributary the parent stream can be followed to its source at the head of the steeply-ascending and steep-sided gully called Lincombe Goyle, which is the principal depression between the ridge covered by South Lincombe Plantation and that of Path Hill, the next in succession of the spurs which project southeastward from the parent ridge of East Hill. The Lincombe brook and its affluents are quite small streams, but sufficient for the lay-out of ornamental waters, and the picturesqueness of the undulations which direct their winding courses helped to determine the selection of this place for the country seat of the owners of Sidbury Manor. The gardens, the fish ponds, the eminence of Evergreen Hill, the "Deer Park" and the adjoining woodland of South Lincombe which rises above, extend for more than a mile from the vicinity of Sidbury village to the open heath land of East Hill.

The Manor of Sidbury, bestowed on the Church by a Saxon King, was held by the Diocese of Exeter until the beginning of the nineteenth century. Since then it has passed through several hands. The present holders of Sidbury Manor have added much land to this estate, including a considerable acreage in Salcombe Regis. The Manor House is late Victorian and its grounds in their present state are modern. The former is of no architectural interest, but the formation of a large park is a valuable addition to the scenery of the Sidmouth countryside. The English park is one of the characteristic attractions of our country, and it is of particular interest to understand the circumstances of its origin. These are well explained in the historical analysis of the subject given in

Professor Patrick Abercrombie's manual of *Town and Country Planning*,[1] from which the following is a condensed quotation:

"During the seventeenth and eighteenth centuries a profound change came over the face of the country: the diversion of wealth from the Church to the aristocracy led to the formation of the great landed estates and the creation of the country houses which abound in England as they do in no other country in the world. The Englishman when he founded a family built a palace in the country but was content with a square box in town: the country house required a garden for its immediate setting and this soon led to the desire for a Park—a piece of enclosed country in which grazing alone was allowed and in which quasi-wild animals—the three types of deer and even wild cattle—might be seen roaming at large. The park added a greater richness to the landscape.... Every country squire became an amateur in planting and in the study of the picturesque. It is still by no means realized—if it can ever be definitely known—how much of England was consciously laid out during this time, far beyond the boundary of the home park.... The regularizing and angular hand of man has thus been softened into heightening an effect here, opening a prospect there, or planting out an unseemly object."

In the endeavour to carry Professor Abercrombie's analysis yet further I am led to enquire—what was the influence which determined all the great landowners in

[1] A volume of the *Home University Library of Modern Knowledge* (1933), see pp. 184 *et seq.*

England to adopt the same type of lay-out for their parks? Abercrombie rightly refers to the influence of Kent the architect, and Capability Brown the landscape gardener, but, was there not some type of natural, or at least pre-existing, landscape which inspired these makers of modern fashion? Judging from my own impressions I should say that the prototype was the "open forest" such as one sees in parts of the New Forest, and also in parts of Sherwood Forest now included in the immense private estates of the Dukeries. It is also relevant to note that in Trinidad, Panama and other tropical countries there are savannahs which remind one strongly of English parks. It seems therefore that the typical English park is the artistic reproduction of a form of landscape which in some places arises from natural growths upon undulating ground and in others through the foresters' activities carried on without thought of appearance.

Just above Sidbury village, Roncombe, a valley 2 miles long, diverges from the eastern side of Sid Vale. Roncombe stream, at times a trickle at other times a torrent, courses down the combe, its ultimate source the clay cap of Broad Down. Not far from the junction of Roncombe with the Sid is the house of Sand Barton, generally known simply as Sand, still the property of the ancient family of Huysh, which has been maintained in its structural amenities, outside and within, from Tudor times to the present day and is thus unique among the considerable mansions of the Sidmouth district.

Its landscape setting is no less attractive than that of Sidbury Manor House but of quite a different kind. In

Tudor times there were no parks in Sid Vale, and from the charming gardens of Sand we step straight into the ordinary farm fields. But Roncombe, more than 3 miles from the sea front and not leading to any crossing place of the plateau, remains entirely rural, and thus the outlook from Sand rivals in scenic charm that of a mansion overlooking a home park. The winding combe, the banks and hedgerows, woods and copses, and steep slopes rising to the wild plateau, with the changing colour of the seasons and soft skies of the Devon coast—what more can any Englishman want for the surroundings of his home?

But a cloud of anxiety hangs over us until a guarantee or covenant ensures that such surroundings shall remain unspoilt. Roncombe Valley provides an opportunity for those concerned with the preservation of English scenery to do their duty, not merely for the advantage of the owners of the ancient house but for the sake of all who enjoy what is left of rural England as it was before the Mechanical Age set in.

§ 12. The west side of Sid Vale above Sidbury village is furrowed by numerous combes directed south-of-east. Only tiny brooks flow down these combes, but much of the water supply of Sidmouth is obtained from the springs hereabout. The eastern ridge of Upper Sid Vale, a projection from the plateau which terminates at Pen Hill a mile to the north of Sidbury village, faces the Sid with a continuous slope 2 miles long with no furrowing combes. Thus the upper part of Sidbury displays clearly the distinctive character of the river system of the

Sidmouth district, throughout which the main streams follow the southerly slope of the plateau and the lesser streams the eastward dip of the Red Keuper Marl which underlies the Greensand. It is at or near the interface of these strata that the tributary streams take their rise.

The appearance now presented by the relief of the Sidmouth countryside stimulates the imagination to picture the place as it was when the streams began to cut out the combes and goyles which have now attained maturity of sculptured form. Pictures thus conjured up provide no small part of the aesthetic enjoyment of scenery. To the best of my understanding the following is the process by which rain and river began to carve out Sid Vale, Salcombe Valley and Weston Combe after the upraising of the Blackdown Plateau. The clay cap must have been much thicker than it is now, and so the water mostly collected in surface streams which followed the general slope of the plateau, which is mainly to the south. The streams flowing in consequence of the general slope (technically termed "consequent" streams) thus determined the direction of the principal valleys. They received very little water by lateral drainage until they had cut down to the surface of the Red Marl. Then "subsequent" streams began. Had the interface between Red Marl and Greensand been horizontal, the tributary streams would have joined the parent stream both on the right bank and the left, but the interface of the strata sloping downwards from the west, the tributaries are almost all from the western side.

Salcombe Valley and Weston Combe have no lateral

goyles except in the immediate neighbourhood of the Combe Head. The contour lines of the sea floor convey a suggestion as to the characterof Salcombe Valley and Weston Combe many thousand years since. The 10, 20 and 30-fathom lines are shown on the bathy-orographical map of the British Isles in *The Times* atlas. The 30-fathom line bears no relation to the shore line of Lyme Bay, but the 20-fathom line running close to Portland Bill and Downend Point (Scabbacombe Head) and embayed between them is evidently connected with the relative rates of erosion of the harder and softer rocks of the bay. In the offing of Salcombe Regis the distance of the 20-fathom line from the present shore is about 12 statute miles. When the sea front was there or thereabout we may picture Salcombe Valley and Weston Combe as the head waters of considerable streams such as those which now flow down Sid Vale and Roncombe Valley.

We now return, after this excursion in geology, to the scenery of Sidbury Parish. The road from Sidbury to Honiton runs on the east side of the Sid and affords an attractive view of the wooded combes which furrow the west side of Upper Sid Vale. It is free from the hoardings for advertisement which frequently line the approaches to seaside resorts. The roads of the Sidmouth district are fairly free from these disfigurements and it is to be hoped that the Local Authority will keep them so. Reaching the plateau at the height of 780 feet the road runs on with very slight ascent and arrives in three-quarters of a mile at the boundary of Sidbury Parish and Sidmouth Urban District. Here at the Hare and Hounds Inn, the road to

Honiton is crossed by that from Ottery St Mary which traverses the plateau on the way to Colyford, following a line almost coincident with the boundary between the Sidmouth district and the parishes of Gittisham, Honiton, Farway and Southleigh as far as Lovehaye Common, where it turns from south to south-east, entering Southleigh parish and joining the road from Sidmouth Gap to Colyford after entering the parish of Branscombe.

The open plateau across which the road to Honiton runs after leaving Sidbury is situated in Gittisham Parish and is called Gittisham Hill. It should be carefully noted that the descriptive word "hill" (which is given on the Ordnance Survey map) does not apply to the *ascent* from Gittisham village but to the flat summit or plateau. Following the Ottery-Colyford road to the eastward we come to that part of the plateau adjacent to Sidbury which belongs to Farway Parish. This is traditionally "Farway Hill", and is so marked upon the map. The name is one which the modern wayfarer would expect to apply to the steep ascent from Farway village. It is interesting to note that the difference between this dialectical use of the word "hill" and the general use of the word to-day is comparable to the different use of the word "alp" by the Swiss of the plains and those who dwell in the upper valleys. To the former "the Alps" mean the high snowy mountains, to the latter the word "alp" applies exclusively to the summer pastures.[1]

The fact that the dialectical use of the word "hill" is common to East Devon and West Somerset illustrates

[1] W. A. B. Coolidge, article "Alp" in *Ency. Brit.* (11th edition).

both the regional geography and the history of these areas. The conquest of England by the Saxons whilst still a heathen people did not extend beyond East Somerset. Here they were held up at the line of the Parret Marshes, where in later times Alfred held up the Danes. The next advance came after the Saxons had been converted to Christianity, and when they swept over West Somerset and the adjacent fertile area of East Devon the Celtic inhabitants were not massacred or enslaved but received as fellow-subjects of the Wessex Kings. This sequence of events helps to explain the circumstance that in the matter of dialect West Somerset is in some ways linked with East Devon.

In mediaeval times, when the open hill was a grazing ground where the right of pasture was held by parishioners in common, it was important for the avoidance of disputes to fix parish boundaries in a readily recognizable position, preferably one determined by some natural obstruction. Between Gittisham Hill and Farway Hill the furrow of Blannicombe runs up to the plateau. Down this combe runs a little stream which joins the Otter at Honiton. Its source is the swamp called Ring-in-the-Mire situated near the head of Blannicombe, and this is the terminal of four parishes, Gittisham, Honiton (to which Blannicombe belongs), Farway and Sidbury.

Ancient tumuli are to be seen on the surrounding plateau, reminding us that this ground was the habitation of our Celtic predecessors. Now that the agricultural population has migrated to the valleys Gypsies make their home on Farway Hill, where gorse and heather grow.

CHAPTER IV

SALCOMBE REGIS

§ 13. Following the narrow ridge south-eastward for about a mile and a half beyond Ring-in-the-Mire we come to Roncombe Corner near the headwaters of Roncombe stream, where the plateau spreads out in Broad Down. A third of a mile farther on, adjoining Mincombe Posts Plantation, is another meeting-place of four parishes, Sidbury, Farway, Southleigh and Salcombe Regis. The boundary of Salcombe Regis and Southleigh runs for 2 miles nearly due south as far as the reputed Roman road to Colyford. The parallel boundary of Sidbury and Salcombe Regis is only half a mile distant. The narrow strip of plateau between these boundaries which has the distinctive name of Chelson was not included in Salcombe Regis at the time of Domesday, being then part of the East Devon forest, but later it was released from forest law and added first to Sidbury and finally to the parish of Salcombe Regis.[1] Excepting this narrow strip of plateau, Salcombe Regis consists entirely of Dunscombe Hill and Salcombe Hill, in the modern sense which includes the sides as well as the summits of the two ridges which terminate in sea cliffs. Thus, as was pointed out in a previous chapter, Salcombe

[1] See J. Y. A. Morshead, "A History of Salcombe Regis", *Trans. Devon. Assoc.* 1898.

Regis comprises the west slope of Weston Combe, the whole of Salcombe Valley (where "Church Town" hamlet lies) and the west slope of Salcombe Hill, thus sharing with Sidmouth parish the seaward end of Sid Vale.

The River Sid from the outflow where it seeps through the shingle up to the junction of the Woolbrook nearly a mile inland is the boundary between the civil parishes of Sidmouth and Salcombe Regis. The Sid, being here controlled by weirs and other restraints, adheres to a permanent course. This was not always so within historic times, and the wanderings of the current may account for the fact that strips of land on the western side were, until recently, included in Salcombe Regis.

In old days this little river was of great economic importance both to Sidmouth and Salcombe Regis, for it was their principal source of mechanical power. This applied to the whole of Salcombe Regis, for the tiny brooks of Weston Combe and Salcombe Valley were not able to turn a mill. Thus the corn had to be brought over Dunscombe Hill and Salcombe Hill and down the farm road called Milltown Lane to the hamlet of Sid where the mill, belonging to the Manor, was situated. The Snodbrook in Paccombe Bottom may, however, have been of sufficient power to turn a mill, as is indicated by the following story:[1]

"Once upon a time some adventurous tenant under the Manor planned a mill in Salcombe parish to be

[1] *The Story of the River Sid*, a paper read by the Rev. James G. Cornish at the annual meeting of the Sid Vale Association, March 28th, 1927.

driven by the stream from Knowle, but the Dean and Chapter of Exeter came down on him and ordered the work to cease. Why? Because they, the holders of the Manor, had long before expended capital in establishing a parish mill not far from Sidcliffe House. The tenants of the Manor were bound to bring their corn to be ground here. So the Bostell mill was abolished."

In the eastern part of Sidmouth town Mill Street and Mill Lane lead down to the mill on the west side of the ford, beyond which is Milford Road in Salcombe Regis. This was the last survivor of the mills. The splash of water and the turning of the wheel was one of the delights of my boyhood when I came from Suffolk, a county of slow streams where we saw no mill wheels, for it was the wind which drove the mills that ground the corn we grew. Milford Road follows the east bank of the Sid for about two hundred yards up stream. Here it is traversed by the road which crosses from Sidmouth to Salcombe Regis by the bridge, built about 1817, which provided a more convenient connection between Sidmouth and Salcombe Regis than that by the Mill Ford.

The pedestrian who has reached the upper end of Milford Road can continue along the east bank of the Sid by the footpath through "the Byes" which faces him on the other side of the Salcombe Road bridge. This footpath, which follows the northward course of the Sid for half a mile to Lyme Park Ford, has for a long time past provided the townspeople of Sidmouth with a promenade alternative to the seaside parade. To visitors who think of Sidmouth simply as their seaside resort the footpath by the Sid is relatively unimportant, but it is otherwise with those who live all the year round by the sea. On a sea front the instantaneous response of sea to sky gives unfailing harmony of tone and colour throughout every change of sunlight and of weather. Even the sea gulls flying to and fro are in harmony with the tone and colour of wave and cloud. But it has been well said that there is monotony in ceaseless change, and those who look only on the sea miss the longer rhythm of response which the countryside makes to the round of the seasons—the flowers, fragrance and song of birds in spring, the deep green shade of summer and the soothing hum of bees, the golden glow of autumn and the tracery of bare boughs against the evening light at the close of winter's day. All this the townspeople of Sidmouth can enjoy in the neighbouring footpath of the Byes.

The first field through which the Byes footpath leads is called The Lawn. This was purchased by the Sidmouth Urban District Council for preservation as a public open space, and some of the fields beyond have been generously bestowed on the National Trust for the same purpose. Both of these preserved areas are part of "the Byes", a name which includes all the meadows near the Sid from the Salcombe Road bridge as far as the village street of Sidford, a mile and a half further inland.

Above Lyme Park Ford, which is near the hamlet of Sid and Milltown Lane, the River Sid is joined on its right, western, bank by the Woolbrook, which is here the boundary of Sidmouth and Sidbury. Just above its junction with the Sid the Woolbrook is joined on its left, north, bank by a tiny stream which is the boundary between Sidbury and Salcombe Regis.

The channel of the Sid beyond Lyme Park Ford changes to east-of-north and is no longer followed by a footpath. The Byes path continuing its direct northern course from Lyme Park Ford onwards, under the name of Byes lane, now follows closely the left, east, bank of the tiny stream already mentioned. It affords a pleasant, quiet, stroll, but is out of sight of the river and has not the special charm and beauty of the Byes footpath. It is much to be desired that the Sidmouth Council should arrange for the opening of a public footpath from Lyme Park Ford along the left, west, bank of the Sid through the meadows opposite the houses of Sidcliffe, Sid Abbey, Champs Farm and Higher Griggs. The River Sid, in its course through these meadows of the Byes, is extremely attractive. The first part runs close under the ridge which was chosen as site of the early villas called Sidcliffe and Sid Abbey, and here the stream in its steep descent is controlled by numerous weirs.[1] The murmur of the little cascades soothes the ear, the flash and foam of falling water delight the eye, and in the deep pools trout may be seen. In the fields of Champs Farm and Higher Griggs we come to the only reach of the Sid which exhibits in full measure the characteristic windings through alluvial gravel which are so marked a feature in the course of the

---

[1] The 100-foot contour line crosses the stream at one and a half miles from the sea.

River Axe as seen from the Southern Railway on the route from London to Sidmouth Junction. A winding ribbon of bright water amidst green fields is always a pleasing feature of the landscape, but this general aspect is not the sole contribution which the sinuous flow provides to the beauty of the scene. Standing on the bank and watching the flow of the stream we see the water running smooth and deep below the concave bank, circling in foaming spirals on the lee of the opposite, convex, bank, and thereafter rippling in its rapid course over stony shallows.

The Byes are so essential a part of the surroundings of Sidmouth that the name comes naturally to the inhabitants, but to the enquiring visitor the word is of an intriguing character. It does not occur in the ordinary dictionary but will be found in J. Wright's *English Dialect Dictionary*. It is referred to the dialect of Somerset, which, as I have already pointed out, is in the western part of the county closely linked with the dialect of East Devon. The reference is as follows:

"*Byes*. sb.pl. Som. (baiz). 1. The corners and ends of a field which cannot be reached by the plough, and must be dug by the hand."

In the early days, when dialects took shape, "the field" meant the cultivated tract of manor or parish, not, as now, a small parcel enclosed with bank or hedge.

When we look at the scarps which show the former river banks cut out by the Sid during its wanderings in the Byes, and also note the signs of recent excursions of the stream, it is not difficult to realize that the whole of the Byes was once hardly more suitable for tillage than the ham or marsh below the Salcombe Road bridge. It seems probable also that one of the Hides of Salcombe Regis mentioned in Domesday, variously spelt Biside, Bisyde and Byside, derives its name from the circumstance that it abuts on the Byes. This Hide is shown on Morshead's map as occupying the fertile marl land of Salcombe Hill between Salcombe Hill Road and Milltown Lane and from the Sidford Road up to the Greensand. The Byes may have been a common where the cattle grazed.

§ 14. The Byes terminate on the north at the reputed Roman road which here forms the boundary between Sidbury and Salcombe Regis. In the open view of the valley numerous elm trees of great height are very decorative features.

Following the road to the east we ascend by the Trow Hill road to the summit of the long ridge which according to modern custom is all included in the term "Salcombe Hill". Here we find the upland plain wholly cultivated and divided in small fields, many of them arable.[1] Thus the scene is very different from the open heath of Peak Hill. The difference of use appears to be largely due to a sufficiency of fertile clay on Salcombe and Dunscombe Hills as contrasted with a gravelly surface on the western watershed of the Sid. High banks

---

[1] In Sid Vale on the contrary most of those fields which were arable in the eighteen-seventies have been turned down to grass, which has deprived the valley of much of its former rich colour, due to red soil.

divide the fields and no distant prospect is open to the eye. The luxuriant hedgerows provide pleasing features, field crops are always interesting, and the quietude of agricultural surroundings is a welcome change from the bustle and noise of a mechanized world. All this, however, is no more than can be found in many parts of rural England, but south of the Roman road we come in rather less than half a mile to one of the most exquisite views in the whole of the Sidmouth countryside, that of Salcombe Valley with the sea beyond.

This combe is V-shaped both in ground plan and vertical section. The symmetry of form enhances the picturesque effect derived from height of hill and rich colouring of vegetation, and the ancient church nestling beneath the converging heights near the head of the combe adds architectural beauty to the natural charm of the valley. Above the church and the hamlet of Church Town the ancient farmhouse of Thorn, and the Thorn tree from which it takes its name, add historic interest to the scene. All these precious features are but a mile or so from the frequented seaside town of Sidmouth, yet so completely shut off by the bold eminence of Salcombe Hill that the peaceful valley seems remote from urban life. This central combe of the parish of Salcombe Regis presents a combination of historic interest with natural beauty which makes it a shrine for the pilgrims of scenery. Vineycroft Lane, a sideway from the Roman road, is the route by which the monks of Exeter came to the old house of Thorn where they held

THE SALCOMBE REGIS THORN

the court of the Manor of Salcombe Regis bestowed on them by the Saxon Kings.[1]

Sixty yards from the house on the way to Vineycroft Lane stands the Salcombe Regis Thorn from which the house, the farm, and a wood belonging to the same estate all take their name. The tree marks the boundary between the cultivated Combe of the Royal demesne and the common grazing ground of the "Hill". A large field near by on the plateau, enclosed from the Common in Cromwellian times, is still known as Thorn Hill. The present Thorn tree is of small size, for it was planted only a few years since to replace the former tree which had died off. Such has been the practice since Saxon times. Even as late as Victorian days there was still a feeling in the parish that the welfare of the inhabitants was in some mysterious way linked with the life of the Thorn. When the tree died, the next was planted with public ceremony and to the accompaniment of music. This was

[1] It is considered probable that the manor of Salcombe Regis was part of the gifts presented to the Monastery of St Peter's, Exeter by Aethelstan in A.D. 924, but according to J. Y. A. Morshead, "actual records fail until Canute (A.D. 1019) was by way of penance restoring all the monastery lands Sweyn had plundered to Ahelwold (the Abbot). Risdon thinks Salcombe was then first granted, but in an old list of Chapter deeds that king gives Stoke Canon, though Athelstan had before. Ours probably was a similar reconveyance" (*Trans. Devon. Assoc.* 1898, vol. XXX, pp. 132-46). This charter is preserved in the Archives of Exeter Cathedral.

The house, now occupied by the tenant of the farm, has features of many different periods, the most ancient being a tiny window in the attic hollowed from a single block of stone and presenting the appearance of late Norman times. This old manor-court house is situated where the 500-foot contour lines of Salcombe and Dunscombe Hills meet in a point at the head of the combe. It stands upon the plateau clay, but the underlying Greensand crops out near by at the foot of the adjoining farmyard. Inland from the farmhouse the rise of the plateau is very gentle in slope.

within the recollection of my friend the late J. Y. A. Morshead and is recorded by him in his paper on Salcombe Regis to which I have so often to refer. At that time the village choir had portable instruments, but when my elder brother James George Cornish, then owner of the entailed Thorn estate, planted the present tree the village choir were tied to a church organ, and the Thorn was planted without ceremony and music. Few of the visitors to Sidmouth as they pass by this small tree have any idea of its historic interest, and so greatly has local lore faded with the substitution of school teaching for home tradition that many people in the parish do not realize that the history of Salcombe Regis, no less than that of Glastonbury, is linked to a Hawthorn tree.

In my opinion it is important for those of us who devote ourselves to the preservation of the scenery of rural England to do what we can to kindle the historic sense as well as to stimulate the feeling for natural beauty. With this in mind, and as owner of the Thorn estate, I have put a monumental stone beside the Salcombe Regis Thorn with an inscription describing its origin.

§ 15. We now return to the scenery of Thorn Farm as it is to-day. Vineycroft Lane is continued by a road leading past the Thorn tree and downhill to the church. On the east side of the road, opposite to the farmhouse, is the meadow called Long Stone, situated on the crest of the plateau and at the head of the combe. This commands a clear view of the V-shaped valley which slopes down to the coast; the restful horizon of the sea, high in the field of view, spanning the gap between the steep sides of Salcombe and Dunscombe Hills. A public footpath crossing Long Stone meadow makes the view accessible to all. A view down a V-shaped combe extending to the sea has charm for us all. The simple outlook on the sea and sky at the end of the valley allows the eye to dwell upon the bold symmetry of the converging hill sides from which the complexity of a land horizon would distract attention. Between these enclosing arms, the sea, when lit by a low sun, appears in the form and brightness of a pendant of sparkling diamonds. Here the horizon of the sea is exceptionally potent in its restful influence, for in such a view no other level line, whether of breakers or the beach, competes for the attention of the eye. Moreover, the sound of the waves heard from a combe is exceptionally soothing, for it is the perfect cadence of periodic breakers, the sides of the combe shutting out the confused sound of waves falling elsewhere at unrelated times.

The valley terminates in the gully of Salcombe Mouth where a steep and narrow path winds down to the shingle beach 100 feet below. No trace of salt works now remains, but there seems no reason to reject the tradition that Salcombe Regis, formerly written Saltecombe, derives its name from this industry, once common on our coasts.

Turning back and following the valley-way inland for three-quarters of a mile from the head of the gully we reach the church, which stands 400 feet above the sea, an elevation much greater than the casual spectator would suppose from the way it nestles under the curving summit of the high plateau. Those interested in the optics of scenery should note that when turning back to walk up the combe the apparent length of the valley is greater than when looking down towards the sea. This is due to the V-shaped plan, for when looking down the valley the diverging sides counteract the perspective, whereas when looking up the valley the perspective is reinforced by convergence of the hill sides.

Taking our bearings from the church, we see that two-thirds of the circumference of the plateau summit is comprised in the fields of Thorn Farm. Exploring these surrounding fields we find beautiful views looking over Salcombe Valley to the sea, especially from the fields which extend to the south along the crest of Dunscombe Hill whence the coast beyond Torquay is seen. Thus there are many attractive sites for the erection of villas. If, however, we pause to consider not only the outlook but the aspect of such villas we realize that, cutting the skyline of the combe, they would greatly impair, indeed almost destroy, the character and beauty of Salcombe Valley.

It remains to describe the charming outlook in other directions from the commanding plateau of Thorn Farm.

North of the reputed Roman road is Orleigh's Hill, charming in the variety of its rough and wooded slope, whence we view the secluded valleys of Paccombe and Harcombe with their crowning copses where the woodcock come in winter. On the north is Buckton Hill, a bold buttress of the plateau. To the north-west we look across Sid Vale to Sidbury Castle, the ancient earthwork which crowns another buttress.

On the west side of Thorn Farm the fields slope down to the well-grown wood called Thornhill Plantation. This commands a general view of Sid Vale, both down to the sea where Sidmouth lies sheltered between the lofty ridges of Peak Hill and Salcombe Hill, and up to Sidbury with its rural setting of wooded goyle and fertile combe. Facing Thornhill Plantation across the valley is the conspicuous feature of Sidmouth Gap, which affords a glimpse of the world beyond Sid Vale.

In the central fields of the farm, farther back on the plateau near the Thorn tree, buildings would not cut the skyline of the neighbouring valleys, but here our national story is interwoven with the scenery, and the historic sense would be shocked if the Thorn tree which marks a boundary of Saxon times and the house where the manor court was held were cluttered up with modern villas.

Finally, there is a detached portion of Thorn Farm, called Brown Close Orchard, situated at a much lower level, which is part of the sylvan surroundings of Salcombe Regis Church. The fields between the farmyard and the orchard (which adjoin the garden of the vicarage) were subtracted from the original farm to make part of the glebe.

That the architectural focus of the English countryside is the parish church is nowhere more apparent than in Salcombe Valley. The church tower, built of stone quarried from Dunscombe Hill near by, is a fine structure of the Perpendicular period. The impression produced by its strength and symmetry is enhanced by a setting amid the soft textures and informal lines of a pastoral scene. So perfect is the church of Salcombe Regis in the combination of form and setting that it has become a place of pilgrimage throughout the year.

Moreover, the importance of preserving the sylvan surroundings has to do not only with the aspect of the church but with the prospect from the churchyard. Those of us who have stood there in time of family tribulation know that sorrow is softened and hope strengthened by the peaceful beauty of the surrounding scene. So the little orchard opposite the church should not be built upon.

In the zoning for house-building under the scheme put forward by the Town Planning Committee of the Sidmouth Urban District Council great care was taken to prevent·any building which would cut the skyline of Sid Vale and thus impair the view from Sidmouth town. Zoning on the Salcombe Valley side of the watershed was however planned on the opposite principle, the semicircle of the plateau from west to east above and near the church being treated as the proper area for the growth of the hamlet of Church Town into a townlet. In this part of Thorn Farm the fields were zoned for no less than six houses to the acre, those south of the Roman road four to the acre, and outlying fields north of the road and on Dunscombe Hill one to the acre. Thus the erection of more than six hundred houses was sanctioned on the 227 acres of Thorn Farm and Thornhill Plantation.

While appreciating the regard shown by the Planning Committee for the interests of the owner I decided not to take advantage of the scheme, and informed the Sidmouth Council that no part of Thorn Farm would be developed for building during my lifetime and that it would remain entirely agricultural land. My purpose being the preservation of Salcombe Regis Valley as a shrine for the devotees of scenery, it may seem that reservation for my lifetime is an inadequate provision, but in this matter it is important to note that the population curve of England clearly indicates that the recent spate of building cannot continue for more than a few years. There is also another consideration. Thorn Farm cannot, as the law now stands, be scheduled for preservation in perpetuity either as an open space or as agricultural land unless it be disentailed. But the maintenance of family connection with the countryside is an important factor in the preservation both of the rural scenery and the rural life of England. Moreover, I know that my next heir "looks to the rock whence we are hewn" and will do nothing to mar the beauty of Salcombe Regis. So I am content that the farm shall remain entailed.

CHAPTER V

SALCOMBE REGIS (concluded)

§ 16. Following the level road which leads from the Thorn tree past the War Memorial on the way to Sidmouth we come to a tract of gorse and bracken on the north side where the Norman Lockyer Astronomical Observatory stands. The choice of the site, although facilitated by circumstances of ownership and residence, was mainly due to observational advantages of position. Situated on the plateau at the height of 565 feet, the level horizon is unbroken, so that every star can be observed for the maximum time. No other English observatory is better situated in this respect. A plateau is better than a peak for an observatory as a peak causes currents of ascending air and these interfere with telescopic definition. Observation of the diffraction rings of stars proves that the air on Salcombe Hill is remarkably steady, a very important factor in astronomical observation. The surrounding vegetation of gorse and bracken helps to keep the hill top from being over-heated, and the neighbourhood of the sea also contributes to steadiness of air.[1] These important advantages are to some extent countered by the fact that mist and cloud are more frequent than in observatories situated in the east of England. Distance from towns (where tremors are

[1] See *Handbook to the Norman Lockyer Observatory* (2nd edition, 1935).

caused by traffic, lights flame to the sky and the stars are dimmed by dust and smoke) is also a great advantage for an astronomical observatory. In recent years certain building projects caused anxiety to the council of the observatory on account of the risk of smoke, and it is therefore fortunate that the fields on the south, which is the more important side, of the observatory were reserved in perpetuity as agricultural land by agreement between their owner, the late James George Cornish, and the Sidmouth Urban District Council under the Town and County Planning Act of 1932.

From the south side of the road opposite to the observatory a public footpath leads to the cliff. This path, being on the western brow of the plateau, commands a remarkable view. The prospect through Sidmouth Gap was mentioned in the second chapter. The mountain triad of Haytor, Saddle Tor and Rippon Tor are also seen in picturesque isolation above the ridge of Peak Hill.

At the point where the path reaches the cliff a stile on the left hand is the entrance to South Combe Farm. The first field comprises the summit of Salcombe Cliff, the highest escarpment in Devonshire of which I have found record in the maps, charts and publications of the Ordnance Survey and Admiralty.[1]

Of all local statistics none are of more interest to people connected with Sidmouth than the heights of the neighbouring cliffs. In order to set at rest the differences of opinion which are prevalent on the subject I quote

[1] See *ante*, Chap. II, § 5.

from letters which have been kindly sent to me by the Admiralty and the Ordnance Survey Department.

From the Hydrographic Department, Admiralty, December 15th, 1937:

"In reply to your letter of 3rd December I have to inform you that 535 feet is the height of the cliff where the ridge known as Salcombe Hill terminates.

"This height was determined during a hydrographic survey 1851/4 and must be increased by 5 feet to obtain the height above ordnance datum."

It follows from the above that the height of Salcombe Cliff, in the landsman's sense, is 540 feet, for Ordnance Datum is Mean Sea Level whereas the datum used by the Hydrographic Department is the level of High Water.

The letter from the Ordnance Survey Office is as follows:

"14th December, 1937.

"In reply to your letter of the 4th instant, the ground level at the Trigonometrical station on High Peak Hill, 1/2500 plan Devon 94/5, is 516 feet above Mean Sea Level; this value was determined when the district was being contoured in 1889, and has not since been verified.

"We have no record of the altitude of the Trigonometrical station on Peak Hill. The surface height 527 shown on the map (on the footpath about 40 yards N.W. of the Trigonometrical station) is the highest recorded by us on that particular escarpment."

The Admiralty *Channel Pilot*, Part 1, gives the height of Peak Hill as 521 feet above High Water, that is to say 526 feet above Mean Sea Level, taking account of the range of tides in this part of the English Channel. Comparing this with the Ordnance Survey figure, and noting the fact that the summit is almost flat, we may take 526 feet as the height of the escarpment.

Of the range of cliffs east of Salcombe Hill the highest is Coxe's Cliff, in the parish of Branscombe, which, according to the Admiralty Chart, is 515 feet above High Water, that is to say 520 feet above Ordnance Datum. Thus the authoritative figures show that Salcombe Cliff where it terminates in a field of gorse and rough pasture belonging to South Combe farm is higher than any of the neighbouring cliffs.

Peak Hill and High Peak are two of the principal features in the view from the summit of Salcombe Hill, and the respective heights of these eminences are of considerable interest to observers of local scenery. The Trigonometrical station on High Peak is at the edge of the cliff at its highest point so that there is no doubt that Peak Hill cliff is the higher. It is difficult to judge the relative heights when looking from the Sidmouth promenade, and the name High Peak has led to the prevalent assumption that this promontory is the higher. In this connection it should be noted that the traditional name is not High Peak Hill, as the eminence is called by the Ordnance Survey Department and in the Memoirs of the Geological Survey.[1] "High Peak" has no flat summit and "Peak Hill" has, and, as I have already pointed out, the word "hill" in the dialect of East Devon

---

[1] See p. 12, *The Geology of Sidmouth and Lyme Regis* (2nd edition, 1911).

means the open grazing ground of a hill top. Thus High Peak has no "hill" in the local and traditional meaning of the word.

Lastly we may enquire, or speculate, whether the names of these features were given, not by the inhabitants of Sidmouth but by those of Otterton, the parish to which the western slope of both eminences belongs. Looking from that side there is no such marked separation of High Peak and Peak Hill as there appears to be when looking from the Sidmouth side, and Peak Hill, the northern part of this continuous eminence, alone has a plateau which would provide a common for grazing.

Reference has been made to the double slope of the hills which enclose the combes of Sidmouth's countryside, steeper above where the formation is Greensand, gentler below where the formation is Red Marl, a tougher substance where vegetation arrests denudation sooner than on the friable Greensand. The section of Salcombe Cliff below the summit field is somewhat similar in the upper portion which has been formed by weathering. The first drop of about 150 feet is a fairly steep face of Greensand. Then a gentle slope of Red Marl juts out and terminates abruptly in steep crags of Red Marl undercut by the sea. These are moreover deeply furrowed by the land waters which come out at the base of the Greensand, so that the lower half of the cliff is modelled in great bastions. These as seen from the Sidmouth esplanade stand out in bold relief of light and shadow as the sun descends. But their most impressive aspect is that from a boat as one rows along under the cliffs. The jutting ridges then appear as supporting buttresses, for scenic impressions are mainly based on ordinary experience, and cliff forms seem to be architectural. The lower parts of the cliff both on the Sid Vale and Salcombe Valley side have no capping of Greensand and here the Red Marl cliff presents a smooth face.[1]

When standing on the edge of the cliff in the summit field of South Combe the jutting projection hides the steep bluff of the lower part. The visual effect is however in a certain way enhanced, for the eye has a different focus for the projecting edge and for the sea so far below, and this increases the observer's impression of altitude.

After receiving my earliest impressions of altitude at the summit of Salcombe Cliffs I visited the French Riviera in undergraduate days. Here from the mountain slopes above Menton I looked out upon the Mediterranean from heights three or four times greater than that of Salcombe Cliff, but the impressiveness of the Riviera heights was not proportionately greater. The emotion aroused is in great measure due to the height of the sea horizon in the field of vision, an effect produced by the unconscious declension of the eye, and this

[1] The jutting part of the Red Marl used to be cultivated for potatoes, and half an acre is still marked as a field belonging to South Combe Farm on the 25-inch Ordnance map. There used to be good rabbit shooting in the rough cover of this part of the cliff, but the risk of cliff falls is now regarded as too great. A picturesque and entertaining account of a day's sport there is given in *Nights with an Old Gunner* (1897) by my eldest brother the late C. J. Cornish. Foxes make their earths in the Greensand of the cliff which is locally termed "Fox mould". When walking on the hill top one may see a fox crossing the footpath with light and springy step on his way to the shelter of the cliff.

lowering of the line of sight is caused as much by steepness as by altitude. Thus when we stand on the grassy edge of "South Down Field" in South Combe Farm, where Salcombe Cliff at its summit is cut off sharply, as by a knife, the eye is so strongly lured below that the horizon of the sea stands near the top of the picture, giving the impression of a vast watery plain. Nothing breaks the silence of the scene except maybe the cry of jackdaws sheltering in the cliff.

The view of Lyme Bay to the west has greatly changed since we left the Sidmouth esplanade. There the westward view is composed of two parts, the cliffs as far as Otterton Point seen in the colouring proper to their substance, and the distant promontories of Tor Bay and beyond, atmospheric in tone and colour. From the summit of Salcombe Cliff the western coast line is not thus apparently divided, the successive promontories appearing in their true relation as parts of one great bay. The point of sight being now far above the level of Otterton Point, the eastern approach to the estuary of the Exe, marked by Straight Point, comes into view, with Beacon Hill (400 feet) crowning the cliff on the west side of Budleigh Salterton. To the left is the line of heights of Babbacombe Bay with the skerry of Oarstone, 112 feet in height, visible half a mile out at sea. The west side of Tor Bay terminating in Berry Head (186 feet) with Downend Point just beyond are as usual the terminal features.

On the east, Beer Head hides the shore line as far as a point at or near Bridport Harbour. The coast between

Seaton and Lyme Regis which runs north-of-east being thus invisible, no promontories jutting out picturesquely against the sea are added to those seen from the esplanade. Beyond the southward bend of the shore line at Charmouth more of the hills behind the coast come into view, and Black Down (the range behind Abbotsbury which attains about 800 feet) stands much higher above the shore line than as seen from the esplanade. Portland, when visible, also stands higher above the water line.

The height of Salcombe Cliff being 540 feet and the eye-height of the observer 545 feet the distance of the sea horizon is $31\frac{1}{2}$ statute miles. The "Bill", or Point, of Portland, about 30 feet high, the farthest part of the island, is distant $35\frac{1}{2}$ statute miles, so that the whole height of the island, except 9 feet or thereabouts, is visible above the brow of the sea. As both ends of the island are steep, its whole length of 4 miles is seen from Salcombe Cliff.

That part of the Chesil Beach which ties Portland to the mainland is approximately 40 feet in height but, as far as my experience goes, is not visible to the naked eye. This may be on account of the very small angle which its height above the horizon subtends at so great a distance.

On the west side of the bay the sea horizon at $31\frac{1}{2}$ statute miles is beyond Downend Point (Scabbacombe Head), so that we see the whole height of that terminal promontory.

The view from the cliff summit of Salcombe Hill is very inspiring to the poetic imagination. Among the early visitors to the new watering place of Sidmouth was John Keble, the poet of devout mind. He used to stay with my people, and the traditions of their affectionate intercourse are still preserved. Among these is the record of his wanderings by the cliffs of Salcombe Hill which inspired the poem entitled "Mountain Scenery" which is allotted to the Twentieth Sunday after Trinity in *The Christian Year*:

> "Where is Thy favour'd haunt, eternal Voice?
> 'Tis on the Mountain's summit dark and high;
> No sounds of worldly toil ascending there
> Mar the full burst of prayer;
> Lone Nature feels that she may freely breathe,
> And round us and beneath
> Are heard her sacred tones;
> Such sounds as make deep silence in the heart,
> For Thought to take her part."

In the present day the description of Salcombe Hill as a mountain would seem so great an exaggeration that it could only be permitted to those who have taken out a poet's licence, but this was not so a century since when these verses were written. Then the hill ranges of Southern England, as for instance the Chalk Downs of Sussex, were known as mountains. The modern practice of restricting the term to heights of 2000 feet or more seems to have originated from the sport of Alpine climbing.[1] Thus in the early part of the nineteenth century we find that Salcombe Hill is a "mountain" not only in poetry

---

[1] See *Scenery and the Sense of Sight*, by Vaughan Cornish, pp. 77–78.

but prose, for in *The Beauties of Sidmouth* by Edmund Butcher, a contemporary of John Keble, the following passage occurs: "The beautiful vale in which the town stands is bounded on both sides by long lofty mountains."

§ 17. The cliffs of England and their outlook on the ocean are a precious heritage of the Nation's scenery. And now with the increase of motoring and the multiplication of villas and bungalows there is thrust upon us the pressing problem of how to reconcile the access of the public to cliff lands with the preservation of the natural beauty on which the benefit of their scenery depends.[1] It has been my duty to decide how best to reconcile these contending requirements in the case of my farm of South Combe which crowns the cliff of Salcombe Hill.

In early days, the chief concern of town planning at Sidmouth was the construction and maintenance of promenades and shelters by the sea, whereas paths along the wild cliffs required no attention. The conditions have, however, been reversed in recent times, and of the present problems of local planning none is more pressing than that of protecting the natural amenities of the cliff lands.

The full grandeur of the scene from the summit of a cliff depends upon the maintenance of an open space on the landward side to balance the expanse of the sea. Thus, not only is a cliff path required but its users should not

---

[1] See "The Cliff Lands of England and the Preservation of their Amenities", by Vaughan Cornish, *The Geographical Journal*, vol. LXXXVI, Dec. 1935.

be cramped by a high fence or paling on the landward side.

There is another aspect of the problem of enjoyment and preservation which also needs attention. Cliff lands are only perfect in their charm when they present a pastoral scene, downland where the shepherd and his flock are wandering, or meadows where the cattle are at graze, or arable where the ploughman and his team are furrowing the field. These most picturesque of all human occupations cannot be seen to advantage where the summit of the cliff is a public playground with seats and shelters; a show place, accessible by car, with the usual accompaniments of notice-boards and baskets for litter.

The second point relating to the scenic importance of South Combe has to do with the ancient church, three-quarters of a mile from the sea, and its surrounding hamlet seen near the head of the combe. The lower fields of the farm bound the seaward view from the church and village. These fields running back a quarter of a mile from the sea would provide charming sites for seaside villas, and under the Town Planning Scheme I was allowed to build sixteen houses. But, if houses were built, the view from the village of Salcombe Regis and from the precincts of the church would be sadly marred. I therefore decided that the best way to preserve the scenery of South Combe Farm was neither to make the top fields a sophisticated show place, nor to make the lower fields a site for villas, but to constitute the farm an "Open Space", thus preserving the pastoral scene for ever. This is within my compass, as the farm is not entailed, but freehold. In order to make the picture more complete, my brother, the late Rev. James G. Cornish, agreed not to allow building in the two adjoining fields of North Combe Farm.

The provisions of the Town and Country Planning Act of 1932 enable the owner to confer upon the local authority power to enforce this preservation of amenities, and accordingly I have entered into an Agreement with the Sidmouth Urban District Council for this purpose. Subject to this restriction the farm remains the property of myself and my heirs.

In order that the Agreement might ensure that the public should have full benefit from the preservation of the pastoral scene, it was desirable that I should include provisions for the public use of footpaths across the land. The most important is the cliff path, but it is the law that if the existing path be carried away by fall of the cliff, the public loses the right of passage. I have therefore included in the Agreement a provision that the right of way shall continue even if the existing path be destroyed by fall of the cliff. This path, connecting with an already recognized right of way up the valley bottom to the church, leads round the outskirts of the farm. For full enjoyment of the scenery, as well as for the advantage of a shorter route for the pedestrian returning to Sidmouth, a diagonal way across the farm is required. I have therefore agreed to recognize as public rights of way certain paths traversing the farm to which access has been hitherto by courtesy.

And now let me recount, as it were in retrospect, the ordinary ramble over Salcombe Hill by residents of Sidmouth, or visitors from the great towns who stay in that delectable seaside resort. The west side of Salcombe Hill has been climbed by a cliff path leading up from the Sidmouth parade, a path now included in a strip which has been purchased by the Sidmouth Council. Surmounting the stile at the top of the hill we enter the first field of South Combe Farm. We now leave the view of the town behind and not a single human habitation distracts attention from the splendour of the outlook on the sea. Following the cliff path down the east slope of the hill to the gully near the beach and thence up the valley through and beyond the meadows of South Combe and North Combe, where the red Devon cattle graze, we come to the church, and the historic aspect of scenery succeeds to that of natural beauty. There is much of interest in this venerable building which occupies a good deal of our time, and so we take the short way back, the diagonal route across the farm leading directly to the top of the cliff, whence the cliff path descends to Sidmouth.

Pausing for a last look from the lofty summit, we realize that the ancient church is not the only place of worship in Salcombe Regis. The summit of the cliff has an outlook on the one hand upon the first and final industry of man, the cultivation of the fruits of the earth; and on the other upon the ocean, which is the very emblem of eternity, for it is unchanged by time. Thus the summit of Salcombe Cliff is one of the cathedrals of Nature, a cathedral in which indeed no services are held,

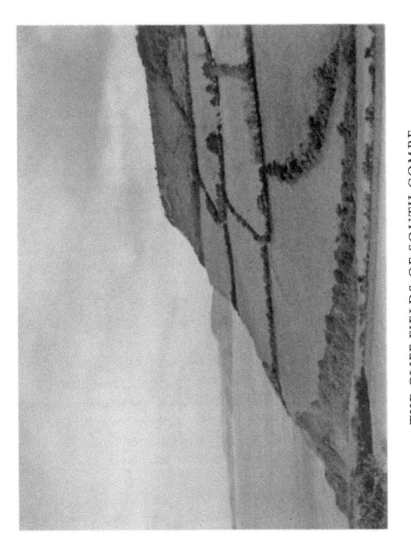

THE CLIFF FIELDS OF SOUTH COMBE

so that we cannot listen to the words of the prophets, but in the silent grandeur of the scene those who enter into communion with Nature can hear the Voice which spoke to the prophets.

§ 18. Facing South Combe Farm on the opposite side of the valley is Coombe Wood Farm which comprises cliff fields on the western slope of Dunscombe Hill. This farm was purchased a few years since by the Sidmouth Council in order to preserve Salcombe Valley from the erection of a hutment camp. The questions which then arose were of special difficulty for those of us who sympathize with the provision of cheap facilities for enjoyment of rural situations by the sea; but, taking all aspects of the matter into account, I am of opinion that the Council were right to adopt a firm attitude, although they had to pay a high price (as was fair and proper) in order to compensate the Company concerned. All those fields of Coombe Wood which face South Combe have been reserved from building by the Council, so that the whole coastal front of Salcombe Valley is now preserved in its rural aspect.

It fortunately happens that there is another combe not far beyond the eastern boundary of the Sidmouth district which is specially suited for camping, preferably in tents or other temporary shelter. This is Branscombe, where there is level access to the beach which Salcombe Valley and Weston Combe have not. It is, however, a narrow opening not sufficient for such town development as has taken place at Sidmouth and Seaton. Branscombe Mouth, now frequented by campers, is the opening of a number of converging combes which twist and turn with a bewildering irregularity which has given rise to the saying that Branscombe was made up of the odds and ends which were left over at the end of Creation, or alternatively that it was "the virst place to be made in the World avore they quite knowed how to do it, lookee!"

Leaving Salcombe Mouth and following the cliff path through Coombe Wood Farm we reach the summit of Higher Dunscombe Cliff, the part of Dunscombe Hill which is visible from the Sidmouth esplanade. In less than half a mile we come to the recession called Lincombe. The sides of this combe go down to the foot of the Greensand slope but only make a slight notch in the Marl below, for the red cliff begins above the 250-foot line. This is Lower Dunscombe Cliff, which is not within the view from Salcombe Hill, Sidmouth esplanade or Peak Hill but is seen from the promontory of Otterton. The footpath, after following the edge of Higher Dunscombe Cliff, turns inland and passes round Lincombe under the brow of the plateau. The rough and wooded slopes of this small combe make a charming foreground for the V-shaped view of the sea, and, the beach being far below the abrupt edge of the cliff, any huts or shelters which may be erected there would be effectually screened. The vulnerable part of the surroundings is the margin of the plateau, and it is essential for the preservation of this little gem of coastal scenery that if houses are built above they should be set so far back that they will not cut the skyline of the combe.

At the cliff end of the combe on its eastern side the path brings us to the foot of a bold feature, the Rempstone Rocks, a vertical crag of irregular front, about 100 feet in height. Between this sheltering crag and the face of Lower Dunscombe Cliff a grassy platform of the Greensand makes a delightful resting place from which to gaze upon the sunlit sea and the far-off headlands of the western bay, their atmospheric softness contrasting with the bold foreground of rough rock.

Neither Peak Hill nor Salcombe Hill has any such crag surmounting the Greensand, for it is on Dunscombe Hill in the parish of Salcombe Regis that for the first time on the way up the English Channel we come to the Chalk rocks, which as we go east become the chief feature of the South Coast. Rempstone Rocks are of the Lower Chalk, the most ancient of the three divisions of the Chalk formation, which lies conformably on the Greensand. Its texture differs considerably from that commonly associated with chalk. When the geologist is dealing with the Lower Chalk of Dunscombe in relation not to its date but its material he describes it as a "shelly limestone".[1] The more familiar kind of chalk is seen at Beer Head, the first of the white cliffs of England and the first occurrence of the Upper Chalk, on the way up Channel.[2]

Although it is in the Rempstone Crag that the outcrop of the Lower Chalk is first conspicuous, it can be traced from the east end of Higher Dunscombe Cliff all the way

---

[1] *Geology of Sidmouth and Lyme Regis* (Mem. Geol. Survey), 2nd edition, p. 59. Here and in *Geology of South-West England* the name "Kempstone" is used, but the Ordnance Survey adheres to the local use of "Rempstone".

[2] *Geology of South-West England*, p. 63.

round the crest of Lincombe, the course of the footpath conforming to it.

The east side of Rempstone Crag faces Weston Cliffs (in Branscombe Parish) on the other side of Weston Mouth, the sea exit of the deep V-shaped valley. The frontier position of this valley is indicated by the circumstance that on the Ordnance map the name of "Dunscombe Bottom" is written on the Salcombe Regis slope and "Weston Combe" on the Branscombe slope. As the valley terminates in "Weston Mouth" it may be best to chose "Weston Combe" as the descriptive name of the whole.

The path which runs along Higher Dunscombe Cliff, round Lincombe and as far as Rempstone Rocks, now turns inland and follows a contour line somewhat above 400 feet for one-third of a mile to the old farmhouse which used to be known as Dunscombe Manor. This path, or trackway, is, so to speak, a terrace between the rough crag of the chalk formation above and Dunscombe Coppice with its fine ash trees on the steep slope of the Greensand below. There are no houses in the valley beneath, and its meadows and woodland with outlook on the sea are a perfect picture of rural England.

To the wayfarer who follows the cliff path westwards from Branscombe, Sid Vale is not visible, only the combe and the distant coast line on the west. Thus Weston Combe, in its calm and solitude, is a shrine for the pilgrim of scenery whether he come from the frequented resort of Seaton or from Sidmouth.

In spring time the Salcombe side of the combe is

carpeted with acres of primroses; in summer, meadow and woodland are a quiet harmony in green; in autumn the coppices light up in golden glow, and in the mild sunshine of a fine winter day the blue waters of the Channel brighten the peaceful valley. This is indeed a scene to be sought not merely for rest and recuperation but for the spiritual inspiration which the beauty of Nature brings.

INDEX

Abbotsbury, 9, 17, 71
Abercrombie, Prof. Patrick, 42, 43
Admiralty, publications, 65, 66
Alps, the, 33
Arthur's Seat, 34
Axe, River, 1, 38, 54

Babbacombe Bay, 70
Bald Hill, 40
Beacon Goyle, 40
Beacon Hill, 32, 34, 35, 37, 39, 40, 70
Beer Head, 1, 2, 20, 27, 28, 70, 79
Berry Head, 18, 70
Bickwell Vale, 15
Blackdown Hills, 1, 31, 71
Blackdown Plateau, 1, 37, 45
Blannicombe, 48
Bostell mill, 51
Bournemouth, 14
Bowd, 35
Branscombe, 2, 25, 29, 47, 67, 77, 78, 80
Branscombe Mouth, 26, 77
Bridport Harbour, 70
Broad Down, 43, 49
Brown Close Orchard, 61
Buckland, Rev. William, 26, 27
Buckton Hill, 31, 60
Budleigh Salterton, 1, 34, 70
Bulverton Hill, 5, 14, 34
Butcher, Rev. Edmund, 4, 5, 6, 12, 73
Byes, the, 51, 52, 53, 54, 55

Camberley, 14
Canopus, 23
Chalk Downs, 72
Champs Farm, 53
Channel Pilot, 18, 66

Charmouth, 71
Chelson, 49
Chesil Beach, 17, 71
Chit Rocks, 3, 10, 24, 30
Church Town, 50, 56, 62
Clayden, A. W., 28, 37 n.
Cliff, definition of, 19 n.; preservation, 73
Clifton Terrace, 24
Coleridge (Samuel Taylor), 12
Coly, River, 38
Colyford, 38, 47, 49
Combe Head, 46
Coombe Wood farm, 77, 78
Connaught Gardens, 24
Coolidge, W. A. B., 47 n.
Core Hill, 34, 35, 39
Cornish, C. J., 69 n.
Cornish, Rev. George James, 15
Cornish, Hubert, 4, 5, 6, 27
  large drawing (1815), 7
  small drawing (1805), 7
Cornish, Rev. James George, 50, 58, 65, 75
Coxe's Cliff, 67

Danger Point, 28
Dartmoor, 31, 32, 33, 36, 37
Downend Point, 17, 18, 31, 46, 70, 71
Dunscombe Bottom, 80
Dunscombe Cliffs, 27
Dunscombe Coppice, 80
Dunscombe Hill, 2, 8, 30, 49, 50, 55, 57 n., 59, 60, 61, 62, 77, 78, 79
Dunscombe Manor, 80

East Hill, 41
East India Company, 13

Edinburgh, 34
Evergreen Hill, 41
Exe, River, 32
  estuary of, 70
Exe Valley, 32
Exeter, 2, 35, 38, 57 n.
  Diocese of, 41
Exmoor, 32, 33, 36, 37

Farway, 47, 48, 49
Farway Hill, 1, 47, 48
Fort Field, 7, 8,
Fortfield Terrace, 7
French Riviera, 13, 69
Freshfield, Douglas, 34

Gittisham, 47
Gittisham Hill, 1, 47, 48
Glastonbury, 58

Ham, the, 4, 5
Harcombe-Sweetcombe Valley, 38
Harcombe Valley, 60
Harpford Wood, 35
Haytor, 33, 65
High Peak, 20, 21, 22, 24, 26, 27, 66, 67, 68
Higher Dunscombe Cliff, 78, 79
Higher Griggs, 53
"Hog's Back Cliffs", 19
Holmes, Sir Charles, 19
Honiton, 38, 46, 47, 48
Hutchinson, Peter Orlando, 11 n., 34 n.

Ilchester, 38

Japan, 14

Keble, John, 15, 72, 73
Kestell-Cornish, Miss, 9

Ladram Bay, 21
Leslie, Robert, 4 n.
Lincombe, 78, 80

Lincombe Goyle, 40, 41
Livonia Ford, 35
"Long Picture," the 8-11
  second edition of, 10
Long Stone, 58, 59
Lovehaye Common, 47
Lower Dunscombe Cliff, 78, 79
Lyme Bay, 46, 70
Lyme Park Ford, 51, 52, 53
Lyme Regis, 1, 26, 31, 38, 71

Mackinder, H. J., 32, 37 n.
Marlborough Place, 7
Menton, 69
Milford Road, 51
Mill Ford, 51
Mill Lane, 51
Mill Street, 51
Milltown Lane, 52, 55
Mincombe Posts Plantation, 49
Monks of Otterton, 7
Morshead, J. Y. A., 49, 55, 57 n., 58
Mutters Moor, 14

Napoleonic Wars, 3
National Trust, 30, 52
New Forest, 43
Newell Arber, E. A., 19
Norman Lockyer Observatory, 18, 35, 64
North Combe Farm, 75, 76

Oarstone, 70
Ordnance Survey Department, 66, 67
Ordnance Survey Maps, 18, 28, 29, 47, 65, 69 n.
Orion, 22, 23
Orleigh's Hill, 60
Otter, River, 1, 31, 35, 36, 48
Otterton, 2, 34, 68
Otterton Point, 1, 2, 20, 21, 22, 23, 28, 70, 78
Otterton Priory, 34
Ottery St Mary, 12, 35, 47

Paccombe Bottom, 50
Paccombe-Chelson Valley, 38
Paccombe Valley, 60
Panama, 43
Panoramic views, 11
Parret Marshes, 48
Path Hill, 41
Peak Hill, 2, 3, 5, 9, 13, 20, 21, 27, 28, 29, 30, 31, 34, 36, 39, 55, 67, 68, 78, 79
Pen Hill, 44
Perpendicular period, 8, 10, 62
Picket Rock, 21
Pinn Hill, 32
Plantagenet times, 3
Pleiades, the, 22
Poole, 36
Portland, Isle of, 1, 17, 22, 31
Portland Bill, 46, 71
Princess Victoria (afterwards Queen), 13
Procyon, 22

Reid, Clement, 28
Rempstone Crag, 79, 80
Rempstone Rocks, 79, 80
Reynolds, Stephen, 4 n.
Ring-in-the-Mire, 48, 49
Rippon Tor, 33, 65
Rocky Mountains, 33
Roncombe Corner, 49
Roncombe stream, 43, 49
Roncombe Valley, 43, 44, 46
Royal Glen, 13

Saddle Tor, 33, 65
St Ives, 11
St Michael's Mount, Normandy, 34
Salcombe Cliff, 25, 65, 66, 67, 68, 69, 70, 71, 76
Salcombe Hill, 2, 8, 9, 13, 18, 19, 20, 21, 26, 28, 30, 36, 39, 49, 50, 55, 56, 57 n., 59, 61, 64, 66, 67, 71, 72, 73, 76, 78, 79
Salcombe House, 25
Salcombe Mouth, 59, 78
Salcombe Regis, 2, 21, 25, 27, 28, 38, 41, 46, 49, 50, 51, 53, 55, 56, 58, 59, 63, 74, 76, 79, 80
  Church, 59, 60, 61, 62
Salcombe Regis Manor, 57 and n.
Salcombe Regis Valley, 2, 45, 46, 50, 56, 60, 61, 62, 63, 69, 77
Saltecombe, 59
Salter's Cross, 14
Sand Barton, 43, 44
Scabbacombe Head, *see* Downend Point
Scientific and Industrial Research, Department of, 31
Seaton, 1, 31, 71, 77, 80
Sea bathing, 6
Sid, hamlet, 52
Sid, River, 1, 2, 3, 4, 35, 38, 39, 40, 43, 44, 46, 50, 51, 52, 53, 54, 55
Sid Abbey, 53
Sid Vale, 2, 15, 31, 35, 36, 38, 39, 40, 43, 44, 45, 46, 50, 55 n., 61, 62, 69, 80
Sidbury, 2, 38, 39, 40, 41, 43, 44, 46, 47, 48, 49, 53, 55, 61
  Church, 38, 40
Sidbury Castle, 39, 40, 61
Sidbury Castle Hill, 40
Sidbury Manor, 39, 40, 41, 43
Sidcliffe, 53
Sidford, 38, 52
Sidmouth, 1, 2, 3, 4, 5, 7, 11, 12, 13, 15, 16, 18, 22, 23, 26, 30, 31, 35, 36, 37, 38, 40, 44, 50, 51, 52, 53, 54, 56, 58, 61, 62, 64, 65, 72, 73, 75, 77, 80
  beach, 9, 23
  Church, 8
  esplanade, 2, 17, 19, 20, 21, 22, 68, 70, 78
  villas, 13
  water supply, 44

Sidmouth Bay, 24
Sidmouth Gap, 2, 34, 35, 36, 37, 38, 39, 47, 61, 65
Sidmouth Manor, 13, 34
  MS. map of, 7, 29
Sidmouth Urban District, 2, 30, 38, 65
  Council, 52, 53, 62, 63, 76, 77
Sirius, 22, 23
Sherwood Forest, 43
Snodbrook, River, 38, 50
South Combe farm, 65, 69, 70, 73, 74, 76, 77, 78
South Lincombe Plantation, 40, 41
Southleigh, 47, 49
Staple Hill, 1
Start Point, 1, 18, 33
Straight Point, 70
"Strawberry Hill Gothic," 13
Sunrise from the sea, 22

Taunton, 31
Thorn farm, 58, 60, 61, 62, 63; reservation as agricultural land, 64
Thorn Hill, 57
Thorn tree, 56, 57, 58, 61, 64

Thornhill Plantation, 61, 62
Tindall, John, 5 n.
Tor Bay, 9, 18, 70
Torquay, 60
Trees, 14
Trinidad, 43
Trow Hill, 31 n., 55

Universal Deluge, 28
Upper Sid Vale, 44, 46
Usher, Bishop, 28

Vineycroft Lane, 56, 57, 58

Wallis' Marine Library, 9, 10
Waves, 23–4
Weston Cliffs, 80
Weston Combe, 2, 45, 46, 50, 77, 80
Weston Mouth, 2, 80
William the Conqueror, 34
Winsford Hill, 36
Woolbrook, River, 35, 50, 52
Woolbrook Church 8
Wright, J., 54

Yarty, River, 1